A STEREOTAXIC ATLAS
OF THE BRAIN OF THE CHICK
(GALLUS DOMESTICUS)

A Stereotaxic Atlas
of the Brain of the Chick
(Gallus domesticus)

WAYNE J. KUENZEL AND MANJU MASSON
Department of Poultry Science
University of Maryland, College Park

THE JOHNS HOPKINS UNIVERSITY PRESS
Baltimore and London

To my mentor and friend, Professor Ari van Tienhoven, Cornell University

Scientific article no. A-4764, contribution no. 7767,
of the Maryland Agricultural Experiment Station (Department of Poultry Science).

The Johns Hopkins University Press, 701 West 40th Street, Baltimore, Maryland 21211
The Johns Hopkins Press Ltd., London

The paper used in this publication meets the minimum requirements of American National
Standard for Information Sciences—Permanence of Paper for Printed Library Materials,
ANSI Z39.48-1984.

Library of Congress Cataloging-in-Publication Data

Kuenzel, Wayne J., 1942–
    A stereotaxic atlas of the brain of the chick (Gallas domesticus).
    Bibliography: p.
    Includes index.
    1. Brain—Anatomy—Atlases. 2. Chickens—Anatomy—Atlases. I. Masson, Manju, 1953–
II. Title.
QL933.K84   1988   598′.617   88-45396
ISBN 0-8018-3700-6 (alk. paper)

Contents

Preface

"The problem of neurology is to understand man himself."
—*Wilder Penfield (1891–1976)*

When one compares the volume of information contributed to neuroscience by experiments using mammalian versus avian species, the former clearly dominates the literature. The stereotaxic atlas of the chick brain is an attempt to encourage greater use of poultry in neurobiological research. The atlas is the first one for an avian species that includes stereotaxic coordinates in three planes of section: cross, sagittal, and horizontal. As explained in the text, there are several published atlases of the avian brain, two of which have served as landmarks. The atlas of the hen telencephalon, diencephalon, and mesencephalon by van Tienhoven and Juhász was the first stereotaxic atlas of an avian species and included considerable detail of hypothalamic structures. The atlas of the pigeon brain by Karten and Hodos was the first avian atlas showing stereotaxic coordinates from the olfactory bulbs to the brainstem in cross and sagittal planes. It also included a surgical technique to position the bill of a bird 45° below the horizontal axis of the plane of a stereotaxic instrument. The important procedure fixed the brain such that its horizontal axis was nearly parallel to the base of the instrument. Hence, the coronal sections of the pigeon brain could therefore be more readily compared with those obtained in the rat, mouse, monkey, and human. More recently, another important contribution was made by an international committee on avian anatomical nomenclature (ICAAN). The committee standardized the nomenclature of all avian anatomical structures and published a book, *Nomina Anatomica Avium*, edited by Baumel.

The present chick atlas includes the important findings made by previous avian brain atlases as well as the nomenclature recommendations of ICAAN. In addition, a simple surgical procedure is described to assist a researcher in adapting the atlas to any chicken (male or female) and any avian species, regardless of age. Each atlas plate was hand drawn from stained preparations. The objective was to produce a plate that resembled a section prepared with both a nissl and a fiber stain. Nuclei presented in each atlas plate appear more distinct than would normally be observed in stained sections or photomicrographs. This was achieved by not including the surrounding neurons that comprise the normal background in slide material. The reason for exaggerating the contrast among the nuclei, fiber tracts, and background was to clarify each structure identified in the atlas. We are fully aware that the procedure has its flaws and we may have omitted or distorted some important structures. We hope that mistakes have been kept to a minimum.

Over the past five years, several individuals were consulted and made very helpful suggestions. We are especially grateful to Professor Sabine Blähser, who critically reviewed the entire atlas and emphasized the importance of adhering to an international nomenclature. Others to whom we are indebted include Professors Ari van Tienhoven, Bill Hodos, Harvey Karten, Tony Reiner, Steve Brauth, and A. Oksche.

The project was supported by funds from the Maryland Agricultural Experiment Station and a Book Subsidy Award granted by the General Research Board of Graduate Studies and Research, University of Maryland, College Park.

A STEREOTAXIC ATLAS
OF THE BRAIN OF THE CHICK
(GALLUS DOMESTICUS)

Introduction

Atlases Available for Avian Species

Stereotaxic atlases of the brains of several different avian species are available to include the chicken, *Gallus domesticus* (van Tienhoven and Juhász, 1962; Yoshikawa, 1968; Feldman et al., 1973; Snapir et al., 1974; Youngren and Phillips, 1978); pigeon, *Columba livia* (Karten and Hodos, 1967); Mallard Duck, *Anas platyrhynchos* L. (Zweers, 1971); Japanese Quail, *Coturnix coturnix japonica* (Bayle, Ramade, and Oliver, 1974); canary, *Serinus canaria* (Stokes, Leonard, and Nottebohm, 1974); Barbary Dove, *Streptopelia risoria* (Vowles, Beazley, and Harwood, 1976); and goose, *Anser anser* (Felix and Kesar, 1982). Among the published atlases, the one of van Tienhoven and Juhász (1962) was the first comprehensive atlas and it remains as one of the most complete with respect to nuclei identified within the preoptic and hypothalamic areas. The atlas of Karten and Hodos (1967) was the first to provide sections through the entire brain. Both histological plates and drawings were included in two planes (coronal and sagittal), and it set the standard for stereotaxic atlases that followed. In particular, it included a convenient means of positioning the head of a bird in a stereotaxic instrument so that the horizontal axis of the brain was parallel to the horizontal axis of the stereotaxic instrument (Karten and Hodos, 1967). The technique included the use of a Revzin adaptor attached to a commercially available stereotaxic instrument. It readily positioned the bill of a pigeon 45° below its normal horizontal attitude when the pigeon is standing erect. The technique has worked well with the canary (Stokes et al., 1974) and it was likewise used for the development of the present atlas.

The Advantages of Two-Week-Old Chicks for Studies of Brain Structure, Development, and Behavior

The advantages of conducting studies on chicks two weeks of age are as follows:

1. The skull is ossified at this time and gives a stable support for positioning the head in a reproducible fashion in a stereotaxic instrument. Skull stability became the overriding reason for the choice of two-week-old chicks. The skull of a day-old chick was found to be quite fragile and difficult to align in a stereotaxic instrument for reproducible, accurate brain surgery. The skull of a one-week-old chick, although certainly sturdier than that of a day-old chick, did not provide nearly the stability of the ossified bone present in chicks at two weeks of age.

1

2. The brain of a two-week-old broiler chick is larger than that of an adult pigeon or a white rat, both of which are frequently used for investigations involving the central nervous system (CNS). Hence at this age, it provides the investigator with a biological preparation of the brain that is on a par with those two commonly used species in terms of the accuracy that can be attained in the stereotaxic location of a structure within the CNS.

3. The body weight of broiler chicks at two weeks of age is about 300 g. A chick with this amount of mass has enough body reserves to withstand any short-term loss of weight and rebound from postoperative effects such as aphagia or adipsia. For this reason, survival rate at this age from surgery is considerably greater than that of chicks at hatch or at one week of age.

4. The chick is considered a true homoiotherm at this age since it has passed the critical period in which body temperature is still labile. A constant body temperature of 41.0° C can be maintained in chicks two weeks old (Freeman, 1965).

5. The blood—brain barrier is thought to develop at three to four weeks of age in chicks (Waelsch, 1955; Lajtha, 1957). Investigators, using two-week-old chicks, can bypass this phenomenon prior to its development as well as test to what degree the barrier is formed at this age.

6. Reports in the literature claim that chicks can be maintained at a physiological age of about 10 days for many months using protein-, amino acid-, or energy-deficient diets just sufficient to fill maintenance requirements (McCance, 1960; Dickerson and McCance, 1960). Return to an unrestricted, nutritional diet restores growth and development to a normal rate with little subsequent effect on adult body size or egg production (McRoberts, 1965). Dietary manipulation may be a convenient means of maintaining chicks of the appropriate size for use with the brain atlas over prolonged periods of time. In addition, the atlas would allow one to test whether the brain likewise can be maintained indefinitely at a physiological age of 10 days.

Atlas Plates Included in the Text

The atlas includes 98 transverse plates at 0.2 mm intervals from the olfactory bulbs to the spinal cord. In addition, there are 25 horizontal and 16 sagittal plates. In contrast to other published atlases of the brain of the chicken, the cross-sectional plates more closely approximate those found in the pigeon and mammalian stereotaxic atlases. The main reason is that similar to the pigeon atlas (Karten and Hodos, 1967), the bill of the chick was positioned at a 45° angle below the horizontal plane of the stereotaxic instrument prior to blocking brain tissue for histology. The atlas is specifically designed for use with the Domestic Chick, *Gallus domesticus* at two weeks of age.

METHODS

Types of Chicks Used for the Atlas

There are two types of chicks readily available that can be purchased for research purposes: broilers and Leghorns. Broiler chicks are derived from stock selected for rapid growth rate and efficient conversion of feed to meat. Leghorns

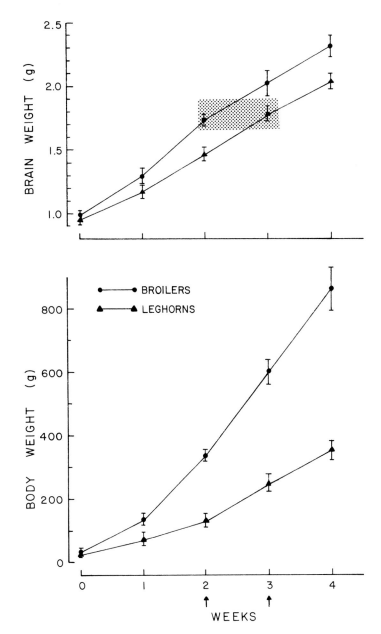

Figure 1. Brain weight and body weight of broiler and Leghorn chicks from day of hatch to four weeks of age; vertical lines mark 2 S.E.M. (*n* = 7 chicks/data point).

are derived from various commercial strains selected for egg production, and hence their growth rate is significantly less than that of broilers.

Figure 1 shows a typical growth curve (total body weight and brain weight) of male broiler and Leghorn chicks. Note that at hatching there is no significant difference in body weight or brain weight between the two types of chicks. At one week of age average broiler weight is double that of Leghorns and is roughly 2.5 to 3.0 times that of Leghorns from two through four weeks of age.

The atlas has been prepared using male broiler chicks two weeks of age. Broilers average 300 to 325 g body weight at this time and have an average brain weight of 1.75 g. If one wishes to use Leghorn chicks with this atlas, it is recommended that chicks of this type be three weeks of age. As can be seen from figure 1, brain weights of the two types are comparable between the two

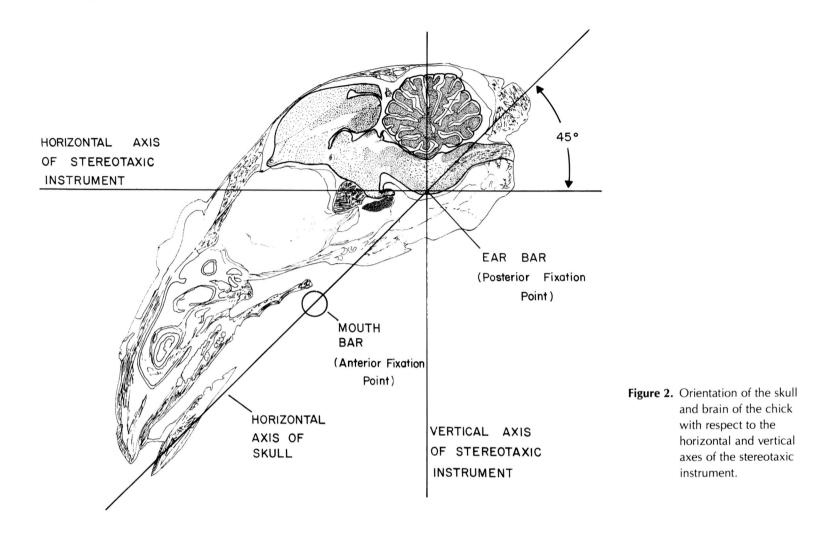

HORIZONTAL AXIS
OF STEREOTAXIC
INSTRUMENT

45°

EAR BAR
(Posterior Fixation
Point)

MOUTH
BAR
(Anterior Fixation
Point)

HORIZONTAL
AXIS OF
SKULL

VERTICAL AXIS
OF STEREOTAXIC
INSTRUMENT

Figure 2. Orientation of the skull and brain of the chick with respect to the horizontal and vertical axes of the stereotaxic instrument.

age groups (see stippling, fig. 1). At three weeks of age, male Leghorn chicks average 225 to 250 g body weight and have a mean brain weight of 1.78 g.

The Orientation of Skull or Head of a Chick to Obtain Atlas Plane

Figure 2 is a schematic representation of the orientation of the skull to make the horizontal axis of the brain parallel to the horizontal axis of the stereotaxic instrument. Two fixation points (anterior and posterior) are required to stabilize the calvarium. The external auditory canals serve as the posterior fixation point. Modified small bird ear bars (altered from a tapered tip of 45° to 36°) are available by special order (Kopf Instruments, Tujunga, California) and provide a means of securing the posterior region of the skull. Fitting the ear bars properly within the canals is the most critical step. It is imperative that the ear bars are directed to the most *posterior* region of the auditory canals. A method found to help plant the ear bars properly is to first insert them loosely in both canals. Then, facing the head of the chick, place a forefinger of each hand at the posterior, ventral corner of the lower mandible and gently move the head forward and dorsally. This should ensure that the tips of the ear bars are in the posterior, ventral region of the auditory canals. Both ear bars should immediately be inserted more deeply into the canals; the distance between the tapered tips in a two-week-old male broiler chick is 11.4 ± 0.45 mm ($\bar{x}$ ± S.E.M; range 10 to 13 mm; $n = 17$).

The anterior fixation point is similar to that described for the pigeon, that is, the caudal, dorsal region of the buccal cavity. Two machined components are required to secure the bill: a pigeon adaptor and a 45° adaptor slide[1] (Kopf Instruments). The pigeon adaptor includes a bar and a beak clamp. The bar is placed in the chick's mouth and gently slid to the rear of the buccal cavity as far as possible *without forcing the corners of the mouth*. The clamp should then be lowered across the upper beak and secured to prevent movement of the upper mandible. Care must be taken to prevent excess pressure of the clamp from distorting the upper beak (see figures 3 and 4 for the type of clamp used).

A final check for proper alignment of the skull in the stereotaxic instrument can be ascertained after an incision of the skin of the head is made and the dorsal area of the calvarium is exposed. One can then look directly down on the skull to determine the position of the bregma (a fissure of the skull parallel to the ear bars formed by the fusion of the frontal and parietal bones) with respect to the ear bars. The bregma should always be anterior to the ear bars, as shown in figure 6B or 6D.

The Preparation of Chick Brains for Use in Developing a Brain Atlas

Nine brains taken from chicks two weeks of age were used for the development of this atlas. Each chick was first anesthetized with an intravenous (IV) injection of Chloropent[2] (1.8 ml/kg). Chicks were then perfused via the heart with 90 ml physiological saline followed by 90 ml Heidenhain's[3] solution. Each head was then positioned in a stereotaxic instrument as described in the previous section and three stainless steel (SS) insect pins (#2) were implanted in each brain at known coordinates. In the case of the brains used to construct the cross-sectional atlas plates, two pins were implanted horizontal to the base of the stereotaxic instrument. Each was inserted into the forebrain and the pins exited either the brainstem or the cerebellum. The third pin was inserted vertical to the base of the stereotaxic instrument. The brains used to construct the sagittal atlas plates had two pins inserted horizontally and one pin inserted vertically to the stereotaxic instrument. The horizontal pins entered the right side of the brain and exited the left side. The brains used for the horizontal plates had two pins inserted vertically and one pin horizontally. The latter entered the forebrain and exited the brainstem.

After the reference pins were inserted into the brains, each brain (still within the skull) was placed in 10% buffered neutral formalin for a minimum of four days. At that time all brains were blocked as described by Karten and Hodos (1967). A #11 SS blade was inserted into an electrode carrier. Each head was properly oriented in a stereotaxic instrument and the three pins were removed. Rongeurs were then used to remove all bone along the dorsal regions of the skull

---

[1]If one wishes to adapt a stereotaxic instrument that is secured to a base plate, e.g., a Model 900 small animal stereotaxic instrument (Kopf Instruments), for use with chicks, it will be necessary to have machined three pedestals to raise the "U" frame at least 12 cm above the steel base plate in order to accommodate the 45° slide.

[2]Each ml contains chloral hydrate, 42.5 mg; magnesium sulfate, 21.2 mg; pentobarbital, 8.86 mg; ethyl alcohol, 14.25%; propylene glycol, 33.8%.

[3]Heidenhain's solution (without mercuric chloride): formaldehyde (40%), 200 ml; glacial acetic acid, 40 ml; trichloroacetic acid, 20 g; sodium chloride, 5 g; distilled water, 800 ml.

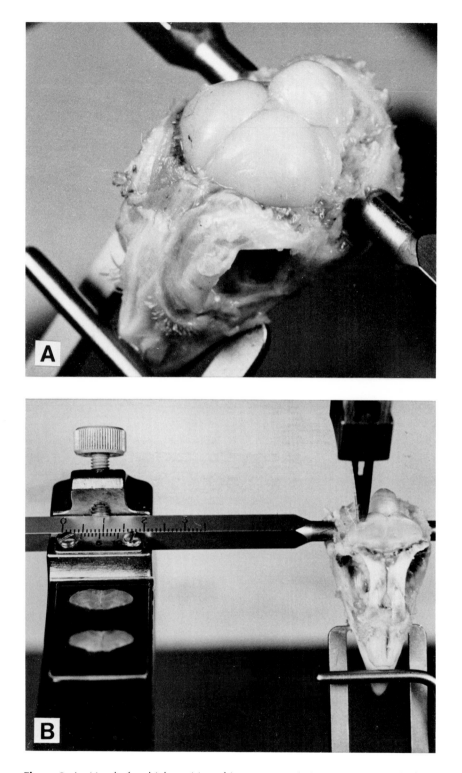

Figure 3. A. Head of a chick positioned in a stereotaxic instrument (SI) with the calvarium removed and the brain exposed.
     B. A scalpel blade fitted in an electrode carrier to block the brain in a vertical plane with respect to the SI. The anterior half of the brain has been removed.

(fig. 3A). The SS blade was then used to sever each brain in half along a known coordinate and in the appropriate plane depending on which set of atlas plates were to be generated by a particular brain (fig. 3B). The two pieces of each brain were then carefully dissected from their respective skulls and placed in 30% sucrose-formalin until each piece sank to the bottom (approximately four days). Brains were then embedded in gelatin-albumin[4] and hardened in a chamber with formaldehyde fumes generated from a 37 to 40% formaldehyde solution. Twenty-four hours later the embedded brain was immersed in 20% buffered formalin overnight, which further hardened the gelatin-albumin matrix. The blocked or cut end of each brain piece was then positioned down on top of a leveled ice pedestal of a sliding microtome.

A SLR 35 mm camera with a 50 mm macro lens was positioned directly above the stage of the sliding microtome (Reichert Scientific Instruments, New York). Brain sections were cut at 40 μm. The unstained flat surface of the embedded brain was photographed every fifth section. A millimeter scale was included in each photograph. Each photographed section was magnified 15X and served as an undistorted outline for each stereotaxic plate. Every fifth and sixth sections were saved for staining with luxol fast blue and cresylecht violet (fifth section) and cresylecht violet (sixth section), respectively (Chroma-Gesellschaft Schmid and Company, distributed by Roboz Surgical Instrument Company, Washington, D.C.).

The stained sections of brain used for each stereotaxic plate were projected onto a surface using a Bausch and Lomb microprojector. They were magnified 15X and manipulated for best fit within each brain outline (obtained from photographs taken of the embedded brains as they were sectioned). All the major nuclei and fiber tracts were subsequently drawn in pencil and then traced in ink. The decision to draw in structures rather than photograph them was made in order to position structures in their estimated, correct location. It also allowed one to highlight certain nuclei and fiber tracts that would not have photographed well. It is hoped that the overall effect of each plate is to delineate clearly most of the known structures of the chick brain yet not overdo the contrast among structures in order to make them approximate what a photographed stained section would have looked like.

The Procedure for Adapting This Atlas to a Bird of Any Age or Size

It is clear that there will be occasions when broiler chicks other than two weeks of age (or Leghorn chicks three weeks of age) will be used for brain surgery. The following is a rapid procedure for obtaining accurate coordinates of a neural structure of a chick of any age or size.

1. Zero an electrode in a stereotaxic instrument (fig. 4A).

2. Correctly orient a fixed brain *in situ* (fig. 4B, refer to previous section "Orientation of Skull or Head of a Chick to Obtain Atlas Plane").

3. Insert a #11 SS blade in an electrode carrier and carefully cut macroslices (0.5 to 1.0 mm thick) of brain tissue (fig. 4B).

4. Remove each brain slice with a spatula and place on a microscopic slide (fig.

---

[4]A two-step procedure:
   (a) Powdered gelatin, 3 g; distilled water, 100 ml; heat until gelatin dissolves.
   (b) Purified albumin, 30 g; stir for about one hour until an even suspension has occurred.

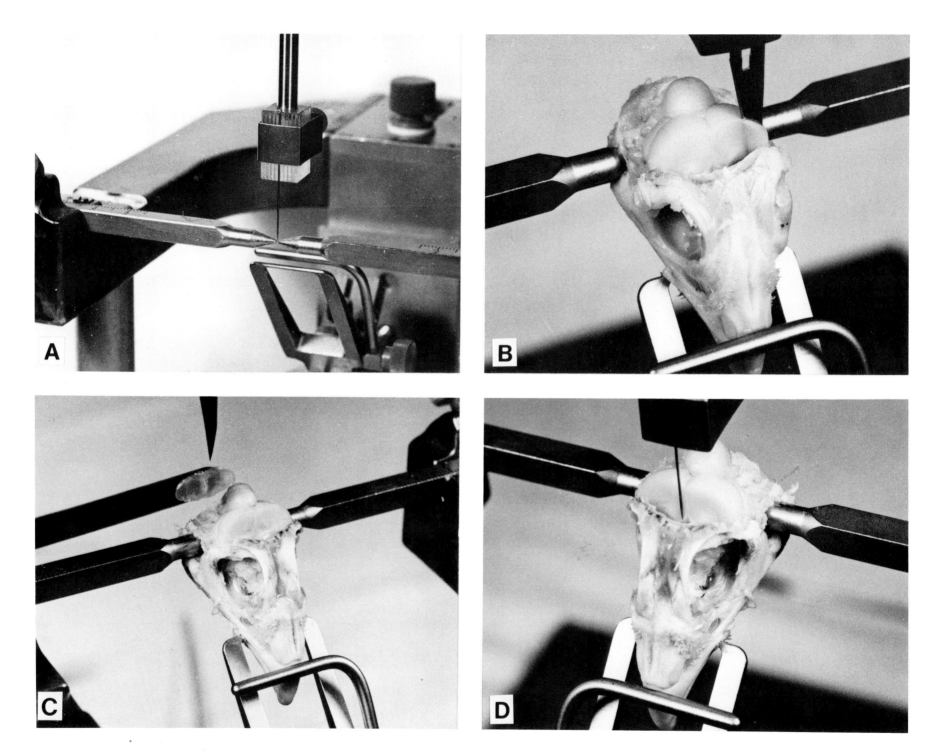

Figure 4. A. An electrode zeroed in a stereotaxic instrument. The tip is positioned directly above the centered ear bars.

B. A scalpel blade fitted in an electrode carrier is used to cut macroslices of brain tissue (0.5 to 1.0 mm thick).

C. A flat spatula is used to remove and transfer macroslices of brain onto glass microscopic slides.

D. The electrode, previously zeroed, is juxtapositioned to an exposed brain locus of interest and the coordinates are then recorded.

4C). To prevent desiccation of the slice, coat with propylene glycol or paraffin oil and cover slip.

5. Continue this process until a neural structure of interest appears in the exposed surface of the brain. For example, in figure 4D the septomesencephalic tract (TSM) is visible in the exposed brain surface. If one were interested in lesioning this tract, one would position the previously zeroed electrode to this structure and read off the three required coordinates (fig. 4D).

6. To obtain coordinates of a structure not so obvious to the naked eye as the TSM, a more refined procedure can be used. For example, if one wished to direct a cannula or electrode to the medial preoptic nucleus (POM) or a specific region of the ectostriatum (E), then a macrosliced brain section can be placed in a photographic enlarger and projected onto a piece of paper. If a permanent record is desired, the unstained brain section can serve as a negative and prints of the image can be made on standard print paper (e.g., Kodabromide F3, F4, or F5 Paper, Eastman Kodak Company, Rochester, New York). Figure 5 A through D gives examples of unstained macroslices of brain printed directly on photographic paper. Groups of cells appear gray, while fiber tracts and myelinated brain areas appear white. In figure 5A one can determine the approximate location of the POM and E (refer to cross-sectional plate A8.8 for identification of these structures). The key, however, is to have a reference point that can be identified both with the naked eye and on the corresponding projected brain section. For example, figure 5B shows a brain slice that is reasonably close to cross-sectional plate A8.2 of the atlas. A neural structure that shows up quite clearly at this brain level is the anterior commissure (CA). It is an ideal structure to serve as a reference point in unstained sections. When slicing macrosections of brain and the CA comes into view, an electrode should then be positioned at the structure and the coordinates recorded. Replace the electrode carrier and electrode with a carrier fitted with a #11 SS scalpel blade. Cut a macroslice of brain and prepare that slice for projection by adding propylene glycol and cover slipping. The unstained projected slice appears as in figure 5B. Next place an electrode or SS insect pin *of known length* on top of the section and project the image using a standard darkroom enlarger (see fig. 5C). One can then readily determine the lateral distance of a structure of interest to midline. The depth coordinate can be determined by first referring to the depth coordinate recorded for the structure observed with the naked eye and identifying that locus on the projected brain slice. All other depth measurements of the brain should be in reference to that locus, which in this example was the CA.

7. By maintaining a library of coordinates of loci within a few different anterior—posterior (AP) planes of a particular-aged chick, one can then make inferences about locations of structures in AP planes not yet prepared by referring to this atlas. Note, however, that two assumptions are made that are probably not completely accurate. One is that all brain areas develop and grow at an equal rate. The second is that the angle formed by the axis of the brain to that of the skull does not change through development. Realizing these limitations, one can in short time determine the appropriate coordinates of several neural structures.

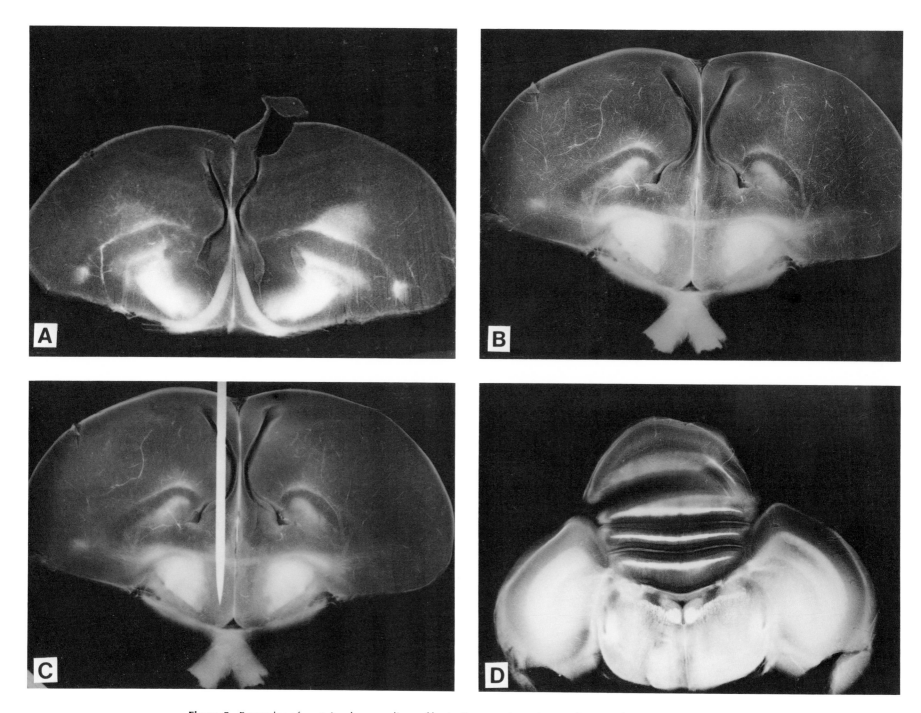

Figure 5. Examples of unstained macroslices of brain tissue projected onto photographic print
paper.
   A. A cross-sectional brain slice of the forebrain comparable to atlas
       plate A8.8.
   B. and **C.** Projection of the same brain slice comparable to atlas plate
       A8.2. In C, an electrode was laid on top of the brain slice.
   D. A section of brain at approximately the level of plate A2.2.

The Procedure for Preparing a Chick for Brain Surgery

The following steps can be taken to insert an electrode at a particular locus within the brain of a chick.

1. Remove feed from chicks about one to three hours prior to surgery.

2. Zero an insulated electrode in the stereotaxic instrument.

3. Anesthetize the chick with one of several methods available (Fedde, 1978). The anesthetic used in our laboratory is Chloropent (see footnote 2 for its preparation). It is administered IV using the brachial vein at a dose of 1.8 ml/kg. A 25 gauge, 15.9 mm needle works well for administering the anesthetic via the wing vein.

4. Place the chick on a support or hammock and carefully insert the ear bars and adjust the clamp for the upper mandible as described in the previous section "Orientation of Skull or Head of a Chick to Obtain Atlas Plane."

5. Shave or remove feathers along the medial region of the head.

6. Wipe the exposed epidermis of the head with isopropyl or 70% ethyl alcohol.

7. Take a pair of surgical scissors and forceps and make a longitudinal cut along the midline of the dorsal skull.

8. Separate the two sides of the skin to expose the calvarium. Remove any blood from the skull and cut edges of dermis using surgical wipes or squares of cheese cloth.

9. Keep the edges of the skin retracted using hemostats as shown in figure 6A. Remove any connective tissue from the dorsal surface of the skull using a bone curette. Remove any additional blood that appeared from the use of the curette.

10. When the skull appears dry, mark the entry point of the electrode on the skull's surface as shown in figure 6A.

11. Remove the electrode carrier and electrode and drill a hole through the calvarium as shown in figure 6B, using a standard dental drill and burr.

12. Attach the electrode carrier again and lower the electrode to the desired depth within the brain.

13. Attach the anode of a lesioning device such as a D. C. Constant Current Lesion Maker (Grass Instruments, Quincy, Massachusetts) to the electrode and the cathode to either the comb or ear bar or within the rectum (fig. 6C).

14. Turn on the current for a particular time period and setting (mA) depending on the size of the lesion desired. For example, a # 1 SS insect pin insulated for use as an electrode, will produce a brain lesion approximately 1 mm^3 when the Lesion Maker is set at 1 mA for 15 seconds.

15. Use a silk suture (4-0) and needle or SS wound clips to close the wound (fig. 6D), after which a topical antibiotic can be used at the incision site. It is recommended that a systemic antibiotic and analgesic be used to aid the chick in recovering from surgery.

16. Remove the chick from the instrument and place it in a heated box (25 to 30° C) or on a heating pad. Check the buccal cavity and swab if necessary to remove excess saliva and other fluids.

Figure 6. Procedure for producing an electrolytic lesion in the brain of a chick.
      A. The skin of the head is cut and the calvarium is exposed in order to mark the entry point of the electrode.
      B. A dental drill and burr is used to make a small opening in the skull.
      C. An insulated electrode is lowered to a brain locus and connected to a constant current lesion maker.
      D. Suturing the epidermis and dermis following neurosurgery.

The Accuracy of the Atlas

Accuracy of the atlas will depend on the type of chick used in experiments; its age, body weight, and sex; and the experience of the investigator. Best results will be obtained when male broiler chicks two weeks of age are used and their body weights are between 250 and 350 g. In our laboratory our best results had success rates of 80% accuracy with the atlas. A more realistic expectation of the accurate placement of electrodes or cannulas by investigators initially using this atlas is 50%.

NOMENCLATURE

The nomenclature found in the current atlas follows that of Karten and Hodos (1967) and Breazile (1979). It should be noted, however, that there is disagreement among neuroanatomists about the names of several avian anatomical structures. Neuroanatomical methods such as golgi and silver stains, horseradish peroxidase, autoradiography and receptor localization, immunocytochemistry and *in situ* hybridization have been used to characterize better specific structures with respect to the material they synthesize, their putative neurotransmitters, receptors, and afferent and efferent pathways. Some authors believe that enough data have been accumulated to suggest avian structures that are homologous to mammalian ones. The following are structures in which some controversy exists with respect to their location, function, or homology.

Anatomical Structures with Controversial Nomenclature

Telencephalon. Table 1 includes the traditional nomenclature of the avian telencephalon in which the suffix -*striatum* is used to describe its subdivisions and a proposed recent terminology appearing in the literature. More data are needed before a determination can be made as to whether homology or homoplasy is the appropriate interpretation of each structure. Nonetheless, it was thought useful to include the table as an aid for those investigators who are more familiar with mammalian neuroanatomy and who are planning to use birds in future studies.

Suprachiasmatic nucleus. Crosby and Woodburne (1940) described the suprachiasmatic nucleus in the dove brain as a small group of cells found in the most rostral plane of the ventromedial hypothalamus. Its corresponding location in the chick brain is labeled SCNm in plates A8.4 and A8.2. Evidence of a retinohypothalamic pathway to the SCNm in the House Sparrow and duck has been provided by Hartwig (1974) and Bons (1976), respectively. More recent studies suggest that there are few, if any, retinal projections to the SCNm and that a small, more laterally situated nucleus is a retinorecipient area (Meier, 1973; Gamlin, Reiner, and Karten, 1982; Cooper, Pickard, and Silver, 1983; Norgren and Silver, 1987; Cassone and Moore, 1987). Hence this more lateral and caudal nucleus has been identified as the avian SCN. Unfortunately, to date no clear evidence has shown that the lateral retinorecipient area serves as a circadian oscillator, as has been demonstrated in the mammalian SCN. Therefore, in the chick brain atlas, this more lateral, caudal nucleus is shown as the ventral nucleus of the supraoptic decussation (DSv, plates A8.0 and A7.8) as originally described by Reperant (1973). Further research is needed to deter-

TABLE 1. Traditional Nomenclature of the Avian Telencephalon and More Recent Terminology Appearing in the Literature

Nomenclature Used in Atlas[1]	Mammalian or Reptilian Structure Having Similar Positions, Afferent and Efferent Pathways, and/or Functions	References

Dorsal Ventricular Ridge (DVR)[2]
Hyperstriatum
  dorsale
  ventrale
Neostriatum
  frontale
  intermedium
  caudale
Ectostriatum
Nucleus basalis
	Isocortex (mammals)
	Anterior DVR (reptiles)[2]
		Northcutt, 1981
		Ulinski, 1983

Archistriatal Complex
Archistriatum
  mediale
  posterior
Nucleus taeniae
	Basal DVR (reptiles)[2]
	
	Amygdala (mammals)
		Ulinski, 1983
		
		Zeier and Karten, 1971

Archistriatum
  anterior
  intermedium
    pars dorsalis
    pars ventralis
	Sensorimotor cortex
	  (primates)
		Zeier and Karten, 1971

Paleostriatal Complex
 	
Large-celled field
  Paleostriatum Primitivum
 	
  Nucleus Intrapeduncularis
Small-celled field
  Paleostriatum Augmentatum
  Lobus parolfactorius
	Corpus Striatum
	  (mammals)
	Basal Ganglia (mammals)
	
	Globus Pallidus
	  (mammals)
	
	Caudate-putamen
	  (mammals)
		Karten, 1969; Karten and Dubbeldam, 1973; Kitt and Brauth, 1981; Reiner, Karten, and Solina, 1983

Nucleus Accumbens
Paleostriatum Ventrale
	Nucleus Accumbens
	  (mammals)
	Substantia Innominata
	  and Ventral Pallidum
	  (mammals)
		Kitt and Brauth, 1981

[1]Nomenclature of Ariëns Kappers, Huber, and Crosby (1936) as modified by Karten and Hodos (1967).

[2]In birds, the DVR includes all structures between the lamina medullaris dorsalis and the lamina frontalis suprema.

mine whether the DSv should be renamed the SCN or SCNl (suprachiasmatic nucleus, pars lateralis).

Inferior olive. On plate P3.2 three of the nuclei comprising the inferior or caudal olivary complex are shown. Vogt-Nilsen (1954) completed a detailed study of the olivary complex and identified at least seven nuclei associated with that structure in birds. It would be of interest to determine the afferent and efferent connections of the olivary complex and compare the structure to its reptilian and mammalian counterparts.

Parabrachial nucleus. Studies are in progress to identify the parabrachial nuclei in the pigeon and determine their afferent and efferent projections (Wild, Arends, and Zeigler, 1987). The ventral parabrachial nucleus is located on plates A1.6 through A1.2.

References

Arends, J. J. A., and H. P. Zeigler. 1986. Anatomical identification of an auditory pathway from a nucleus of the lateral lemniscal system to the frontal telencephalon (nucleus basalis) of the pigeon. *Brain Res.* 398:375–81.

Ariëns Kappers, C. U. A., G. C. Huber, and E. C. Crosby. 1936. The Comparative Anatomy of the Nervous System of Vertebrates, Including Man. Republished in 1967 by Hafner, New York.

Bayle, J. D., F. Ramade, and J. Oliver. 1974. Stereotaxic topography of the brain of the quail. *J. Physiol.* (Paris) 68:219–41.

Bons, N. 1976. Retinohypothalamic pathway in the duck (*Anas platyrhynchos*). *Cell Tissue Res.* 168:343–60.

Boord, R. L. 1968. Ascending projections of the primary cochlear nuclei and nucleus laminaris in the pigeon. *J. Comp. Neurol.* 133:523–42.

Breazile, J. E. 1979. Systema nervosum centrale. *In* Nomina Anatomica Avium (J. J. Baumel, ed.). Academic Press, New York, pp. 417–72.

Cajal, Ramón Y. S. 1911. Histologie du système nerveux de l'homme et des vertébrés (2 vols). A. Maloine, Paris.

Cassone, V. M., and R. Y. Moore. 1987. Retinohypothalamic projection and suprachiasmatic nucleus of the House Sparrow, *Passer domesticus*. *J. Comp. Neurol.* 266:171–82.

Cooper, M. L., G. E. Pickard, and R. Silver. 1983. Retinohypothalamic pathway in the dove demonstrated by anterograde HRP. *Brain Res. Bull.* 10:715–718.

Cowan, W. M., L. Adamson, and T. P. S. Powell. 1961. An experimental study of the avian visual system. *J. Anat.* 95:545–63.

Crosby, E. C., and R. T. Woodburne. 1940. The comparative anatomy of the preoptic area and the hypothalamus. *Res. Publ. Assoc. Res. Nerv. Ment. Dis.* 20:52–169.

Dickerson, J. W. T., and R. A. McCance. 1960. Severe undernutrition in growing and adult animals. III. Avian skeletal muscle. *Br. J. Nutr.* 14:331–38.

Eden, A. R., and M. J. Correia. 1982. An autoradiographic and HRP study of vestibulocollic pathways in the pigeon. *J. Comp. Neurol.* 211:432–40.

Fedde, M. R. 1978. Drugs used for avian anesthesia: a review. *Poultry Sci.* 57:1376–99.

Feldman, S. E., N. Snapir, M. Yasuda, F. Treuting, and S. Lepkovsky. 1973. Physiological and nutritional consequences of brain lesions: a functional atlas of the chicken hypothalamus. *Hilgardia* 41:605–29.

Felix, B., and S. Kesar. 1982. Stereotaxic Atlas of the Goose Diencephale (*Anser anser*). Achevé D'imprimer sur les Presses de L'imprimerie Bialec, Nancy.

Freeman, B. M. 1965. The relationship between oxygen consumption, body temperature and surface area in the hatching and young chick. *Br. Poultry Sci.* 6:67–72.

Gamlin, P. D. R., A. Reiner, and H. J. Karten. 1982. Substance P-containing neurons of the avian suprachiasmatic nucleus project directly to the nucleus of Edinger-Westphal. *Proc. Natl. Acad. Sci. USA* 79:3891–95.

Hartwig, H. G. 1974. Electron microscopic evidence for a retinohypothalamic projection to the suprachiasmatic nucleus of *Passer domesticus*. *Cell Tissue Res*. 153:89–99.

Hunt, S. P., and N. Brecha. 1984. Avian optic tectum: a synthesis of morphology and biochemistry. *In* Comparative Neurology of the Optic Tectum (H. Vanegas, ed.). Plenum Press, New York, pp. 619–48.

Karten, H. J. 1969. The organization of the avian telencephalon and some speculations on the phylogeny of the amniote telencephalon. *Ann. N.Y. Acad. Sci*. 167(1):164–79.

Karten, H. J., and J. L. Dubbeldam. 1973. The organization and projections of the paleostriatal complex in the pigeon (*Columba livia*). *J. Comp. Neurol*. 148:61–90.

Karten, H. J., and W. Hodos. 1967. A Stereotaxic Atlas of the Brain of the Pigeon (*Columba livia*). Johns Hopkins Press, Baltimore.

Kitt, C. A., and S. E. Brauth. 1981. Projections of the paleostriatum upon the midbrain tegmentum in the pigeon. *Neuroscience* 6:1551–66.

Kooy, F. H. 1917. The inferior olive in vertebrates. *Folia Neurobiol*. 10:205–369.

Lajtha, A. 1957. The development of the blood–brain barrier. *J. Neurochem*. 1:216–27.

McCance, R. A. 1960. Severe undernutrition in growing and adult animals. I. Production and general effects. *Br. J. Nutr*. 14:59–73.

McRoberts, M. R. 1965. Growth retardation of day-old chickens and physiological effects at maturity. *J. Nutr*. 87:31–40.

Meier, R. 1973. Autoradiographic evidence for a direct retinohypothalamic projection in the avian brain. *Brain Res*. 53:417–21.

Norgren, R. B., Jr., and R. Silver. 1987. The avian suprachiasmatic nucleus: anatomical and functional aspects (Abstract). *Soc. Neurosci*. 13:211.

Northcutt, R. B. 1981. Evolution of the telencephalon in nonmammals. *Ann. Rev. Neurosci*. 4:301–50.

Nottebohm, F., T. M. Stokes, and C. M. Leonard. 1976. Central control of song in the canary, *Serinus canarius. J. Comp. Neurol*. 165:457–86.

Reiner, A., H. J. Karten, and A. R. Solina. 1983. Substance P: localization within paleostriatal-tegmental pathways in the pigeon. *Neurosci*. 9:61–85.

Reperant, J. 1973. Nouvelles donnees sur les projections visuelles chez le pigeon (*Columba livia*). J. Hirnforsch 14:151–87.

Sanders, E. B. 1929. A consideration of certain bulbar, midbrain and cerebellar centres and fiber tracts in birds. *J. Comp. Neurol*. 49:155–222.

Snapir, N., I. M. Sharon, F. Furuta, S. E. Feldman, S. Lepkovsky, H. Ravona, and B. Robinzon, 1974. An x-ray atlas of the sagittal plane of the chicken diencephalon and its use in the precise localization of brain sites. *Physiol. Behav*. 12:419–24.

Stokes, T. M., C. M. Leonard, and F. Nottebohm. 1974. The telencephalon, diencephalon and mesencephalon of the canary, *Serinus canaria,* in stereotaxic coordinates. *J. Comp. Neurol*. 156:337–74.

Ulinski, P. S. 1983. Dorsal Ventricular Ridge. John Wiley and Sons, New York.

van Tienhoven, A., and L. P. Juhász. 1962. The chicken telencephalon, diencephalon and mesencephalon in stereotaxic coordinates. *J. Comp. Neurol*. 118:185–97.

Vogt-Nilsen, L. 1954. The inferior olive in birds, a comparative morphological study. *J. Comp. Neurol*. 101:447–81.

Vowles, D. M., L. Beazley, and D. H. Harwood. 1976. A stereotaxic atlas of the brain of the Barbary dove, *Streptopelia risoria. In* Neural and Endocrine Aspects of Behavior in Birds (P. Wright, P. G. Caryl, and D. M. Vowles, eds.). Elsevier, Amsterdam, pp. 351–94.

Waelsch, H. 1955. Blood—brain barrier and gas exchange. *In* Biochemistry of the Developing Nervous System (H. Waelsch, ed.). Academic Press, New York, pp. 187–207.

Watanabe, T., N. Iwata, and M. Yasuda. 1975. Further studies on the hypoglossal nucleus in the fowl. *Anat. Hist. Embryol.* 4:323–33.

Wild, J. M., and H. P. Zeigler. 1980. Central representation and somatotopic organization of the jaw muscles within the facial and trigeminal nuclei of the pigeon (*Columba livia*). *J. Comp. Neurol.* 192:175–201.

Wild, J. M., J. J. A. Arends, and H. P. Zeigler. 1987. Efferent projections of the parabrachial nuclei in the pigeon (Abstract). *Soc. Neurosci.* 13:308.

Yoshikawa, R. 1968. Atlas of the Brains of Domestic Animals. University of Tokyo Press, Tokyo; Pennsylvania State University Press, University Park.

Youngren, O. M., and R. E. Phillips. 1978. A stereotaxic atlas of the brain of the three-day-old domestic chick. *J. Comp. Neurol.* 181:567–600.

Zeier, H., and H. J. Karten. 1971. The archistriatum of the pigeon: organization of afferent and efferent connections. *Brain Res.* 31:313–26.

Zweers, G. A. 1971. A Stereotaxic Atlas of the Brainstem of the Mallard (*Anas platyrhynchos L.*). A Stereotaxic Apparatus for Birds and Investigation of the Individual Variability of Some Head Structures. Koninklijke Van Gorcum and Corp. N. V., Assen., The Netherlands.

Index of Structures

This index lists, in alphabetical order of the full name, the structures depicted in the atlas plates. The column at the left displays the abbreviations as they appear on the plates. Below the full name of each structure is a compilation of the transverse, sagittal, and/or horizontal plate(s) on which the structure is shown.

AL   Ansa lenticularis
A8.2–A3.8, L2.2–L1.0, D2.4–D1.6

AQ   Aqueductus mesencephali
A4.0

AA   Archistriatum anterior [rostrale] (Zeier and Karten)
A9.2–A8.2, L5.8, D4.4–D3.4

AId   Archistriatum intermedium, pars dorsalis (Zeier and Karten)
A8.2–A6.0, L7.0–L5.8, D5.2–D4.8

AIv   Archistriatum intermedium, pars ventralis (Zeier and Karten)
A8.0–A6.0, L7.0–L4.6, D4.4–D2.4

Am   Archistriatum mediale (Zeier and Karten)
A7.6–A6.2, L4.2, D4.4–D3.2

Ap   Archistriatum posterior [caudale] (Zeier and Karten)
A5.8–A5.2, L7.8–L7.0, D4.8–D4.0

CDL   Area corticoidea dorsolateralis
A7.2–A4.2, L7.0–L3.8

APH   Area parahippocampalis
A7.4–A4.0, L3.8–L0.2

APa   Area postrema
P2.6–P2.8, L0.6–L0.2, D3.4

AP   Area pretectalis
A5.0

TPO   Area temporo-parieto-occipitalis (Edinger, Wallenberg, and Holmes)
A8.6–A8.0

AVT   Area ventralis (Tsai)
A4.0–A3.4, L1.0–L0.6, D1.6

BCS   Brachium colliculi superioris
A4.0–A2.0, D2.0–D1.6

BC   Brachium conjunctivum
A2.6–P0.2, D4.8–D2.8

BCA   Brachium conjunctivum ascendens
A3.0–A2.6

BCD   Brachium conjunctivum descendens
A2.4–A1.8, D1.6–D1.2

BO   Bulbus olfactorius
A14.8–A13.6, L0.6–L0.2, D3.2–D1.2

CC   Canalis centralis
P3.6–P4.6

CIO   Capsula interna occipitalis
A7.4–A6.8, D7.2–D6.8

Cb   Cerebellum
A5.0–P4.6, L3.8–L0.2, D8.0–D6.0

CO   Chiasma opticum
A8.6–A6.0, L0.6–L0.2, D0.4–D-1.2

Ep   Cingulum periectostriatale (periectostriatal belt)
A11.0–A10.4, L5.8

CPa   Commissura pallii
A8.0–A7.8, D4.0

CA   Commissura anterior [rostralis] (Anterior commissure)
A8.4–A8.2, L3.4–L0.2, D3.6–D3.2

CP   Commissura posterior [caudalis] (Posterior commissure)
A5.0–A4.4, L1.8–L0.2, D5.2–D4.8

CT   Commissura tectalis
A4.8–A2.4, L1.0–L0.2, D5.2–D4.8

CCV   Commissura cerebellaris ventralis
A0.2–P0.6, D5.6

CTz   Corpus trapezoideum (Papez)
A1.8–P1.0, D0.0

CPP   Cortex prepiriformis
A14.6–A14.4, L1.4–L1.0, D3.6

CPi   Cortex piriformis
A8.4–A4.8, L7.8, D5.2–D3.4

DBC   Decussatio brachiorum conjunctivorum
A2.8

D IV   Decussatio nervi trochlearis
A2.2–A1.4, D4.4

DSM   Decussatio supramamillaris
A4.4, D1.6–D1.2

DSD   Decussatio supraoptica dorsalis
A7.4–A6.2, L1.0–L0.2, D1.2–D0.4

DSV   Decussatio supraoptica ventralis
A7.4–A6.4, L1.0–L0.2, D0.8–D0.4

DS   Decussatio supraoptica
A5.8

E   Ectostriatum
A11.8–A8.6, L5.8–L3.0, D7.2–D4.4

ME   Eminentia mediana (Median eminence)
A5.2–A4.4, L0.2, D-1.2

FDB   Fasciculus diagonalis Brocae
A9.2–A8.8, D3.6–D2.4

FLM   Fasciculus longitudinalis medialis
A3.8–P4.0, L0.2, D4.0–D2.0

FPL   Fasciculus prosencephali lateralis (Lateral forebrain bundle)
A9.2–A6.6, L3.8–L1.0, D4.4–D2.4

FPM   Fasciculus prosencephali medialis (Medial forebrain bundle)
A8.0–A7.8

FU   Fasciculus uncinatus (Russell)
A1.6–A0.2

FUm   Fasciculus uncinatus, pars medialis
A0.2

FL   Field L
A7.4–A5.8, D7.2–D6.4

FLO   Flocculus
P0.4–P0.8, D7.2–D6.0

FRL   Formatio reticularis lateralis mesencephali
A4.0–A2.8, L3.0–L2.2

FRM   Formatio reticularis medialis mesencephali
A4.2–A3.4, L1.4

F   Fornix
A8.0

FD   Funiculus dorsalis
P3.8–P4.6, L1.4

FLt   Funiculus lateralis
P4.2–P4.6

FV   Funiculus ventralis
P4.2–P4.6, L0.6–L0.2

P   Glandula pinealis (pineal gland)
A5.8–A5.0, L0.6–L0.2, D8.0–D6.8

Hp   Hippocampus
A8.6–A4.0, L4.6–L0.2, D8.0–D7.2

HA   Hyperstriatum accessorium
A14.8–A7.6, L3.0–L0.2, D8.0–D4.4

HD   Hyperstriatum dorsale
A14.4–A10.0, L1.8–L1.4, D8.0–D4.8

HIS   Hyperstriatum intercalatum supremum
A14.4–A10.2, D8.0–D7.6, D6.0

HV   Hyperstriatum ventrale
A14.8–A5.6, L7.0–L0.6, D8.0–D4.0

HVd   Hyperstriatum ventrale, pars dorsalis
A14.6–A13.6

HVv   Hyperstriatum ventrale, pars ventralis
A14.6–A13.6

LAD   Lamina archistriatalis dorsalis (Zeier and Karten)
A8.6–A5.8, L7.0, D5.2–D2.8

LFS   Lamina frontalis superior
A14.8–A7.6, L3.0–L0.6, D8.0–D4.8

LFSM   Lamina frontalis suprema
A14.8–A8.8, L1.8–L0.2, D8.0–D4.8

LH   Lamina hyperstriatica
A14.6–A5.6, L7.0–L0.2, D8.0–D4.4

LMD   Lamina medullaris dorsalis
A12.4–A6.8, L5.8–L0.2, D6.4–D2.4

LT   Lamina terminalis
L0.2, D2.0–D0.8

LM   Lemniscus medialis
P1.8–P3.6

LS   Lemniscus spinalis
P0.6–P3.6, L2.2–L1.8, D2.0–D0.4

L   Lingula; vinculum lingulae (ICAAN)
A0.8–P1.0, D3.4

LPO   Lobus parolfactorius
A12.4–A9.0, L2.2–L0.2, D3.6–D2.8

LoC   Locus ceruleus
A3.0–A1.0, L1.8–L1.4, D4.0

N   Neostriatum
A14–A5.6, L7.8–L3.0, D8.0–D4.8

NC   Neostriatum caudale
A5.4–A4.4, L4.2–L0.2

NF   Neostriatum frontale
L5.8, L4.6, L2.6–L1.8

NI   Neostriatum intermedium
A9.2–A8.4, L2.6–L1.0, D6.4–D5.6

N IX-X   Nervi glossopharyngeus et vagus
P2.4–P3.6, D4.0

N VI   Nervus abducens
AP0.0–P1.0, D1.2–D0.0

N VII   Nervus facialis
A0.2

N XII   Nervus hypoglossus
P3.8–P4.0, L1.0

N VIII c Nervus octavus, pars cochlearis
P1.4–P1.8, L3.0–L2.6

N VIII v Nervus octavus, pars vestibularis
P0.4–P1.2, L2.2–L1.0, D4.0–D2.8

N III Nervus oculomotorius
A3.8–A3.4, L0.6–L0.2, D2.8–D1.2

N V Nervus trigeminus
A0.8–A0.2, L3.4–L2.6, D2.8–D2.0

N IV Nervus trochlearis
A1.4

N X Nervus vagus
P3.4–P3.6

NH Neurohypophysis
L0.2, D-0.8

GC Nuclei gracilis et cuneatus
P3.6–P4.6

Ac Nucleus accumbens
A10.0–A8.2, L1.0–L0.6, D4.8–D3.2

An Nucleus angularis
P0.8–P1.6, L2.6, D4.8–D4.4

ALA Nucleus ansae lenticularis anterior [rostralis]
A6.8–A6.2, L1.4–L1.0, D2.0

ALP Nucleus ansae lenticularis posterior [caudalis]
A5.2–A4.6, L1.8–L1.4, D2.8–D2.0

AM Nucleus anterior [rostralis] medialis hypothalami
A8.2–A7.4, L0.6–L0.2, D1.6–D0.8

Bas Nucleus basalis
A13.4–A11.6, L3.8–L1.8, D4.4–D2.4

CMOd Nucleus centralis medullae oblongatae, pars dorsalis
P3.4–P4.0

CMOv Nucleus centralis medullae oblongatae, pars ventralis
P3.2–P4.0, D2.8

CS Nucleus centralis superior (Bechterew)
A3.0–A2.6, D2.0–D1.6

Cbl Nucleus cerebellaris internus
A0.8–P0.4, L1.0, D7.2–D6.0

Cblvm Nucleus cerebellaris internus, pars ventromedialis
AP0.0–P0.4, D6.0

CbIM Nucleus cerebellaris intermedius
A0.2–P0.2, L1.4, D6.4–D6.0

CbL Nucleus cerebellaris lateralis
P0.2–P0.6, L2.6–L1.8, D6.4–D6.0

CL Nucleus cervicalis lateralis
P4.0–P4.6, D4.0

nCPa Nucleus commissurae pallii (Bed nucleus pallial commissure)
A8.6–A7.8, L0.2, D4.4–D3.2

Co Nucleus commissuralis (Haller)
P4.4–P4.6

CE Nucleus cuneatus externus (Karten and Hodos) nucleus cuneatus accessorius [lateralis] (ICAAN)
P2.6–P3.6

D Nucleus of Darkschewitsch; nucleus paragrisealis centralis mesencephali (ICAAN)
A5.0–A4.6, D3.6–D3.2

nDBC Nucleus decussationis brachiorum conjunctivorum
A2.8–A2.6

DSv Nucleus decussationis supraopticae, pars ventralis (Reperant); nucleus suprachiasmaticus, pars lateralis (Meier)
A8.0–A7.8

DIP Nucleus dorsointermedius posterior thalami (Karten and Hodos); nucleus dorsointermedialis caudalis (ICAAN)
A5.6–A5.4

DLA Nucleus dorsolateralis anterior [rostralis] thalami
A6.4–A6.2, L3.0, D5.2

DLAmc Nucleus dorsolateralis anterior [rostralis] thalami, pars magnocellularis
A8.0–A7.6, D3.2–D2.8

DLAl Nucleus dorsolateralis anterior [rostralis] thalami, pars lateralis
A7.2–A6.6, L3.4, L2.6–L2.2, D3.6–D2.8

DLAm Nucleus dorsolateralis anterior [rostralis] thalami, pars medialis
A7.2–A6.6, L2.2–L1.4, D4.8–D3.6

DLP Nucleus dorsolateralis posterior [caudalis] thalami
A6.0–A5.2, L2.2, D4.4

DMA Nucleus dorsomedialis anterior [rostralis] thalami
A7.0–A6.2, L1.0–L0.2, D5.2–D4.0

DMN Nucleus dorsomedialis hypothalami
A5.4–A5.0, D0.8

DMP Nucleus dorsomedialis posterior [caudalis] thalami
A6.0–A5.2, D4.4

EW Nucleus of Edinger-Westphal; nucleus nervi oculomotorii, pars accessoria (ICAAN)
A3.4–A2.8, L0.2, D4.4

TD V Nucleus et tractus descendens nervi trigemini
P0.6–P4.6, D3.4, D2.8–D2.4

GLdp Nucleus geniculatus lateralis, pars dorsalis principalis
A6.0–A5.0, L3.0–L2.2, D1.2–D0.8

GLv Nucleus geniculatus lateralis, pars ventralis
A7.8–A5.2, L3.4–L1.4, D1.6–D0.4

HL Nucleus habenularis lateralis
A6.4–A5.4, L0.6, D5.6

HM   Nucleus habenularis medialis
      A7.0–A5.4, L0.6–L0.2, D5.6

IH   Nucleus inferioris hypothalami
      A5.6–A4.8, L0.2, D0.4–D-0.4

IN   Nucleus infundibuli hypothalami
      A5.6–A4.6, L0.2, D-0.8

ICo   Nucleus intercollicularis
      A4.4–A4.0, L4.2–L3.0, D3.4–D3.2

IC   Nucleus intercalatus
      P2.0–P2.6, D3.4–D3.2

ICH   Nucleus intercalatus hypothalami
      A4.4, D1.2

ICT   Nucleus intercalatus thalami
      A7.4–A6.4, L2.2, D2.0–D1.6

IP   Nucleus interpeduncularis
      A3.4–A2.8, D2.0–D0.8

IS   Nucleus interstitialis (Cajal)
      A5.0–A4.6

IPS   Nucleus interstitio-pretecto-subpretectalis
      A4.8–A4.2

nI   Nucleus intramedialis (Huber and Crosby), nucleus c
      (Rendahl)
      A4.8–A4.6, D1.6

INP   Nucleus intrapeduncularis
      A9.8–A8.8, L3.8–L2.6, D4.8–D4.0

Imc   Nucleus isthmi, pars magnocellularis
      A4.2–A1.8, L5.0–L3.4, D5.6–D1.6

Ipc   Nucleus isthmi, pars parvocellularis
      A3.8–A2.0, L5.0–L2.6, D5.2–D2.4

IO   Nucleus isthmo-opticus
      A2.4–A1.8, L2.2–L1.4, D4.8–D4.4

La   Nucleus laminaris
      P0.4–P1.0, L1.8–L1.0, D4.4–D4.0

LA   Nucleus lateralis anterior [rostralis] thalami
      A8.2–A7.4, D2.4–D2.0

LLd   Nucleus lemnisci lateralis, pars dorsalis (Groebbels)
      A2.4–A1.8

LLi   Nucleus lemnisci lateralis, pars intermedia (Arends &
      Zeigler); nucleus lemnisci lateralis, pars lateroventralis
      (Boord); nucleus ventralis lemnisci lateralis (Karten and
      Hodos)
      A1.8–A1.0, L2.6, D2.8–D2.0

LLv   Nucleus lemnisci lateralis, pars ventralis (Groebbels)
      A1.8–A1.6, D2.4

LMmc Nucleus lentiformis mesencephali, pars magnocellularis
      A6.4–A5.6, L3.8, D2.4–D1.6

LMpc Nucleus lentiformis mesencephali, pars parvocellularis
      A6.0–A5.6, D2.4–D1.6

LC   Nucleus linearis caudalis
      A2.4–A1.2, D2.4–D1.2

MCC   Nucleus magnocellularis cochlearis
      P0.8–P2.0, L1.8–L0.6, D4.4–D4.0

MPOd Nucleus magnocellularis preopticus (van Tienhoven),
      pars dorsalis
      A8.8

MPOm pars medialis
      A8.8

MPOv pars ventralis
      A8.8

ML   Nucleus mamillaris lateralis
      A4.6–A4.4

MM   Nucleus mamillaris medialis
      A4.8–A4.4, L0.2, D0.4–D0.0

MLd   Nucleus mesencephalicus lateralis, pars dorsalis
      A4.0–A2.8, L4.2–L3.0, D5.6–D4.0

nVM   Nucleus mesencephalicus nervi trigemini
      A4.6–A3.4, L0.6–L0.2, D5.2–D4.8

MPv   Nucleus mesencephalicus profundus, pars ventralis
      (Jungherr)
      A4.0–A3.2, D2.4–D1.6

MnX   Nucleus motorius dorsalis nervi vagi
      P2.0–P3.8, L0.2, D4.4–D3.6

Mn VII d   Nucleus motorius nervi facialis, pars dorsalis
      A0.4–P0.4, L1.8, L1.0–L0.6, D2.8–D2.0

Mn VII i   Nucleus motorius nervi facialis, pars intermedia
      P0.4

Mn VII v   Nucleus motorius nervi facialis, pars ventralis
      A0.2–P0.2, D1.6–D1.2

Mn V   Nucleus motorius nervi trigemini
      A0.8–AP0.0, D3.2–D2.0

n VI   Nucleus nervi abducentis
      P0.2–P1.2, L1.0, D2.8–D2.4

n XI   Nucleus nervi accessorii (Spinal accessory nerve [Eden
      and Correia])
      P4.6

n IX   Nucleus nervi glossopharyngei
      P1.6–P2.8, D4.0–D3.6

n IX-X   Nucleus nervi glossopharyngei et nucleus motorius
      dorsalis nervi vagi
      P2.0–P2.8

n XII   Nucleus nervi hypoglossi (Nottebohm, Stokes, and
      Leonard), pars tracheo- syringealis, pars lingualis;
      nucleus nervi cervicalis medialis (Watanabe, Iwata, and
      Yasuda)
      P2.6–P4.2, L1.0–L0.2, D4.4, D3.6

OMd   Nucleus nervi oculomotorii, pars dorsalis
      A2.8–A2.4, L0.2

OMdl Nucleus nervi oculomotorii, pars dorsolateralis
A3.4–A3.0, D4.0

OMdm Nucleus nervi oculomotorii, pars dorsomedialis
A3.4–A3.0, D4.0

OMv Nucleus nervi oculomotorii, pars ventralis
A3.4–A2.4, D3.6–D3.2

n IV Nucleus nervi trochlearis
A2.4–A1.6, L0.2, D4.0–D3.4

OA Nucleus olfactorius anterior [rostralis]
A14.2–A13.6, D3.4–D3.2

OI Nucleus olivaris inferior (Kooy and Vogt-Nilsen);
complexus olivaris caudalis (ICAAN); components of OI
include: OAD, OAM, and OP
P2.8–P4.0, L1.8–L0.2, D3.2–D1.2

OI-OAD Nucleus olivaris accessorius dorsalis
P3.2

OI-OAM Nucleus olivaris accessorius medialis
P3.2

OI-OP Nucleus olivaris principalis
P3.2

OS Nucleus olivaris superior
P0.2–P1.2, L2.2–L1.8, D2.4–D1.6

nBOR Nucleus opticus basalis; nucleus ectomamillaris (nucleus
of the basal optic root)
A4.8–A3.4, L1.8–L1.0, D1.2–D0.8

OV Nucleus ovoidalis
A6.4–A5.4, L1.0–L0.6, D4.0–D3.4

Pap Nucleus papillioformis
A3.2–A1.8, L1.4–L0.6, D1.6–D0.4

PBv Nucleus parabrachialis, pars ventralis
A1.6–A1.2

PMI Nucleus paramedianus internus thalami
A5.4–A5.2, D3.6–D3.2

PVN Nucleus paraventricularis magnocellularis
(Paraventricular nucleus)
A8.0–A6.4, L0.6–L0.2, D3.2–D2.4

PHN Nucleus periventricularis hypothalami
A5.8–A5.0, D1.6–D0.8

PL Nucleus pontis lateralis
A2.4–P0.8, L1.8, D1.6–D0.4

PM Nucleus pontis medialis
A1.6–P0.8, L1.4–L0.6, D0.4–D0.0

PV Nucleus posteroventralis thalami (Kuhlenbeck)
A6.4

PMM Nucleus premamillaris
A4.6

POM Nucleus preopticus medialis (van Tienhoven)
A9.0–A8.8, L0.6–L0.2, D2.4–D1.6

POD Nucleus preopticus dorsolateralis
A9.2

POP Nucleus preopticus periventricularis
A8.8–A8.4

PT Nucleus pretectalis
A5.4–A4.8, L3.4–L2.6, D4.0–D3.2

PTD Nucleus pretectalis diffusus
A6.0–A5.8

PTM Nucleus pretectalis medialis
A5.6–A5.2

PPC Nucleus principalis precommissuralis
A6.6–A5.6, L3.4–L3.0, D2.4–D1.6

R Nucleus raphes
A1.6–P3.4, D2.4–D0.0

Rgc Nucleus reticularis gigantocellularis
P1.0–P2.8, L0.6, D2.4

RL Nucleus reticularis lateralis
P3.0–P4.0, D2.0

Rpgl Nucleus reticularis paragiganto-cellularis lateralis (ICAAN);
nucleus paragigantocellularis lateralis (Karten and Hodos)
P1.6–P2.8, D1.6–D1.2

RPaM Nucleus reticularis paramedianus (ICAAN); nucleus
paramedianus (Karten and Hodos)
A1.6–P2.4, D1.2–D0.4

Rpc Nucleus reticularis parvocellularis
P1.4–P2.8, L1.8–L1.4, D1.6–D1.2

RP Nucleus reticularis pontis caudalis
AP0.0–P1.2

RPgc Nucleus reticularis pontis caudalis, pars gigantocellularis
A1.8–A0.2, L1.8–L1.4, L0.6–L0.2, D2.4–D1.2

RPO Nucleus reticularis pontis oralis
A2.6–A1.6, L1.0, D2.4–D1.6

RST Nucleus reticularis subtrigeminalis
P1.8–P3.6, D3.6–D2.0

RSd Nucleus reticularis superior, pars dorsalis
A8.0–A6.8, L1.8–L1.0, D3.4–D2.4

RSv Nucleus reticularis superior, pars ventralis
A7.8–A7.4, L1.8–L1.4, D2.0

ROT Nucleus rotundus
A7.2–A5.6, L3.0–L1.8, D3.2–D1.6

Ru Nucleus ruber
A4.4–A3.4, L1.0–L0.2, D3.2–D2.0

SLu Nucleus semilunaris
A2.6–A1.6, D4.0–D2.8

nPr V Nucleus sensorius principalis nervi trigemini
A1.0–A0.4, L2.6–L1.8, D4.0–D2.8

SL Nucleus septalis lateralis
A9.4–A7.6, L1.0–L0.6, D5.6–D4.4

SM   Nucleus septalis medialis
     A8.6–A7.6, L0.6, D4.8–D4.4

SpL Nucleus spiriformis lateralis
     A5.2–A4.4, L3.0–L2.2, D4.0–D2.4

SpM Nucleus spiriformis medialis
     A5.2–A4.4, L2.2–L1.8, D4.0

nST Nucleus striae terminalis (Bed nucleus, stria terminalis)
     A7.6, D4.8

SCd Nucleus subceruleus dorsalis
     A2.4–A1.4, D2.8

SCv Nucleus subceruleus ventralis
     A2.6–A0.2, D3.2–D2.8

SHL Nucleus subhabenularis lateralis
     A5.8–A5.4

SHM Nucleus subhabenularis medialis
     A5.8–A5.6

SP   Nucleus subpretectalis
     A5.2–A4.2, L3.0–L2.6, D2.0–D1.2

SRt Nucleus subrotundus
     A6.4–A6.0, L1.0–L0.6, D3.2–D2.8

SCNm Nucleus suprachiasmaticus, pars medialis
     A8.4–A8.2

SOe Nucleus supraopticus (Ralph), pars externus
     A8.8

SOv Nucleus supraopticus (Ralph), pars ventralis
     A9.2–A8.8, D2.4–D1.6

SS   Nucleus supraspinalis (Wild and Zeigler)
     P2.8–P4.6, L1.0–L0.2, D3.6–D3.2

Tn   Nucleus taeniae
     A8.2–A6.2, L4.6–L3.8, D4.4–D2.8

Ta   Nucleus tangentialis (Cajal)
     P0.6–P1.2, D3.4

TD   Nucleus tegmenti dorsalis (Gudden)
     A1.4–A1.2, D3.4–D3.2

TV   Nucleus tegmenti ventralis (Gudden)
     A2.4–A1.6, D3.2–D2.4

TPc Nucleus tegmenti pedunculo-pontinus, pars compacta
     (Substantia nigra)
     A3.6–A2.8, L2.2–L1.8, D3.2–D2.0

nTSM Nucleus tractus septomesencephalicus (Nucleus
     superficialis parvocellularis)
     A7.0–A5.2, L2.2–L1.4, D4.4

S   Nucleus tractus solitarii
     P2.0–P4.2, L1.0–L0.6, D4.8–D4.4

T   Nucleus triangularis
     A6.4–A5.6, L2.2–L1.8, D3.6–D3.2

VLT Nucleus ventrolateralis thalami
     A8.4–A7.8, D1.6–D1.2

VMN Nucleus ventromedialis hypothalami
     A7.2–A5.8, L0.6–L0.2, D1.2–D0.0

VeD Nucleus vestibularis descendens
     P0.8–P2.8, L1.8–L1.0, D3.6–D3.2

VeDL Nucleus vestibularis dorsolateralis (Sanders)
     P0.4–P0.6

VeL Nucleus vestibularis lateralis
     AP0.0–P1.2, D4.8–D3.2

VeM Nucleus vestibularis medialis
     A1.0–P2.2, L0.6–L0.2, D4.4–D3.2

VeS Nucleus vestibularis superior
     A0.4–P0.2, L2.6–L2.2, D4.8–D4.0

PVO Organum paraventriculare (Paraventricular organ)
     A6.0–A4.6, D2.0–D0.8

LSO Organum septi laterale (Lateral septal organ)
     A9.6–A7.8, L0.6–L0.2, D5.2–D3.6

SCO Organum subcommissurale (Subcommissural organ)
     A5.2–A4.2, L0.2, D5.2–D4.8

SSO Organum subseptale (Subseptal organ [Legait and Legait;
     subfornical organ in brains having a fornix]) Organum
     interventriculare (Interventricular organ [Blähser])
     A8.4–A7.6, D4.4–D3.2

STO Organum subtrochleare (Subtrochlear organ)
     A2.2–A1.8, D4.0

OVLT Organum vasculosum lamina terminalis
     A8.6

PA   Paleostriatum augmentatum (Caudate putamen)
     A11.8–A6.8, L5.8–L2.2, D6.4–D3.6

PP   Paleostriatum primitivum (Globus pallidus)
     A10.2–A7.6, L5.0–L2.6, D6.0–D4.0

PVT Paleostriatum ventrale (Kitt and Brauth)
     A9.2–A8.0

PCVL Plexus choroideus ventriculi lateralis (Choroid plexus
     within lateral ventricle)
     A7.6–A6.2

PCV III Plexus choroideus ventriculi tertii (Choroid plexus within
     third ventricle)
     A6.2–A6.0, D6.0

PH   Plexus of Horsley
     P1.6–P3.2

PLCV Processus lateralis cerebello-vestibularis
     AP0.0, D5.6

Rx V M Radix mesencephalica nervi trigemini
     A2.8–A2.4

LHy Regio lateralis hypothalami (Lateral hypothalamic area)
     A8.2–A4.8, D1.2–D0.4

RI   Recessus inframamillaris; Recessus infundibuli
(Infundibular recess)
A4.4–A4.2, D-0.4–D-0.8

RPR   Recessus preopticus
D0.8

*SAC   Stratum album centrale
A5.4–A1.6, L5.8–L2.2, D6.8–D1.2

SCE   Stratum cellulare externum
A6.0–A4.6, L1.0–L0.2, D2.8–D2.0

*SGC   Stratum griseum centrale
A6.0–A1.0, L7.0–L2.2, D6.8–D0.8

*SGFS   Stratum griseum et fibrosum superficiale
A6.0–A0.8, L7.0–L1.8, D7.6–D0.0

*SO   Stratum opticum
A6.0–A0.8, L7.0–L3.4, D5.6–D0.0

SMe   Stria medullaris
A7.0–A6.0, L1.0, D5.6–D4.4

SG   Substantia gelatinosa Rolandi (trigemini)
P3.4–P4.6

GCt   Substantia grisea centralis
A5.0–A1.6, L1.4, L0.6, D4.8–D4.4

*SGP   Stratum griseum periventriculare
A5.2–A2.2, L5.0–L2.6, D6.4–D2.0

TeO   Tectum opticum, colliculus mesencephali; tectum
mesencephali (ICAAN)
A6.6–A6.2

ToS   Torus semicircularis
A4.0

CHCS   Tractus cortico-habenularis et cortico-septalis
A7.4–A6.8, D5.6–D5.2

CH   Tractus corticohabenularis
A5.8

DA   Tractus dorso-archistriaticus
A7.2–A5.4

FA   Tractus fronto-archistriaticus
A11.8–A9.2, L5.8, D4.4–D4.0

FT   Tractus frontothalamicus et tractus thalamofrontalis
A9.8

HIP   Tractus habenulointerpeduncularis
A5.8–A5.6

IF   Tractus infundibularis
A5.0–A4.6

TIC   Tractus isthmocerebellaris
A2.6–A2.0, D2.0

TIO   Tractus isthmo-opticus
A6.6–A2.0, L3.0–L1.8, D5.6–D4.0

LO   Tractus lamino-olivaris
P0.2

TnBOR   Tractus nuclei optici basalis (Tractus nuclei
ectomamillaris; tract of the basal optic root)
A5.0

TOV   Tractus nuclei ovoidalis
A5.8–A5.6, D3.2–D2.4

OM   Tractus occipitomesencephalicus
A8.0–A2.4, L5.0–L4.2, L1.8–L1.0, D4.0–D2.8

TrO   Tractus opticus
A7.2–A5.6, L3.8–L1.0, D1.2–D0.0

PST   Tractus pretecto-subpretectalis
A5.0–A4.8

QF   Tractus quintofrontalis
A11.8–A2.0, D1.2

TSM   Tractus septomesencephalicus
A9.4–A5.0, L2.6–L0.2, D5.2–D2.0

TS   Tractus solitarius
P2.0–P3.6

SCbd   Tractus spinocerebellaris dorsalis
P1.4–P2.4, D3.2–D2.4

TT   Tractus tectothalamicus
A6.4–A5.4, L2.2, D2.0–D1.2

TTS   Tractus thalamostriaticus
A6.4–A6.0, D4.0–D3.2

TVM   Tractus vestibulomesencephalicus (Papez)
A4.2–A3.0, L1.0, D4.4

TO   Tuberculum olfactorium
A11.4–A9.4, L1.4–L0.2, D2.8–D2.4

Va   Vallecula telencephali
A14.2–A9.4, D8.0–D5.6

VC   Ventriculus cerebelli
AP0.0–P0.4, D6.8–D6.0

VL   Ventriculus lateralis
A13.2–A7.4, L5.0–L0.2, D8.0–D2.8

VO   Ventriculus olfactorius
A14.8–A13.4, D2.4–D1.6

V IV   Ventriculus quartus (Fourth ventricle)
P1.0

VT   Ventriculus tecti mesencephali
A4.8–A2.4, L5.8–L1.4, D6.4–D2.0

V III   Ventriculus tertius (Third ventricle)
A8.6–A7.6, D5.6–D0.0

*Details of systems of nomenclature for layers of the optic tectum can be
found in Supplementary Plate A4.6, ENLARGEMENT OF OPTIC TECTUM,
p. 81.

Plates

The number used to identify each plate and all coordinates given along the axes of plates signify mm.

Transverse Plates: A 14.8 through P 4.6 where distance is A (Anterior) or P (Posterior) from a zero reference plate AP 0.0 (p. 104).

Sagittal Plates: L 7.8 through L 0.2 where L = Lateral distance from midline (L 0.0).

Horizontal Plates: D 8.0 through D -1.2 where D = Depth within brain tissue from a zero reference plate marking the level of the entrance of the earbars into the auditory canals (D 0.0, p. 165).

BO   Bulbus olfactorius
HA   Hyperstriatum accessorium
HV   Hyperstriatum ventrale
LFS   Lamina frontalis superior
LFSM Lamina frontalis suprema
VO   Ventriculus olfactorius

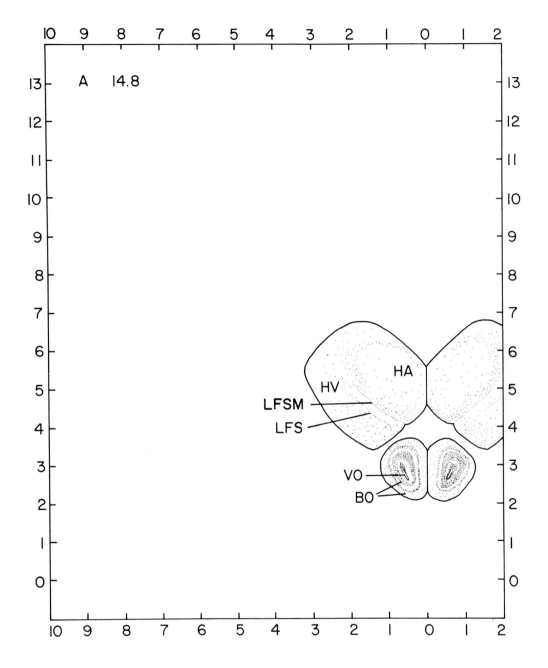

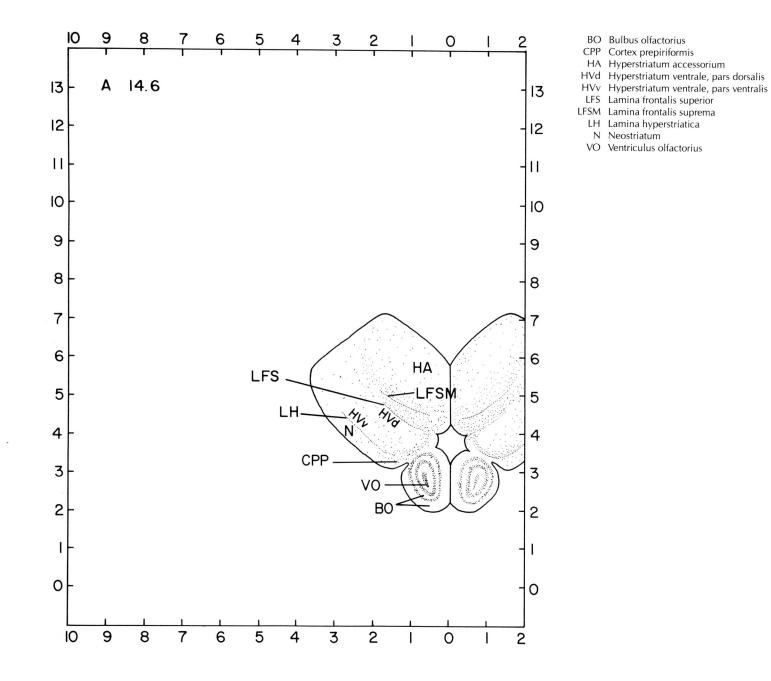

BO   Bulbus olfactorius
CPP   Cortex prepiriformis
HA   Hyperstriatum accessorium
HVd   Hyperstriatum ventrale, pars dorsalis
HVv   Hyperstriatum ventrale, pars ventralis
LFS   Lamina frontalis superior
LFSM Lamina frontalis suprema
LH   Lamina hyperstriatica
N   Neostriatum
VO   Ventriculus olfactorius

BO   Bulbus olfactorius
CPP   Cortex prepiriformis
HA   Hyperstriatum accessorium
HD   Hyperstriatum dorsale
HIS   Hyperstriatum intercalatum supremum
HV   Hyperstriatum ventrale
HVd   Hyperstriatum ventrale, pars dorsalis
HVv   Hyperstriatum ventrale, pars ventralis
LFS   Lamina frontalis superior
LFSM Lamina frontalis suprema
LH   Lamina hyperstriatica
N   Neostriatum
VO   Ventriculus olfactorius

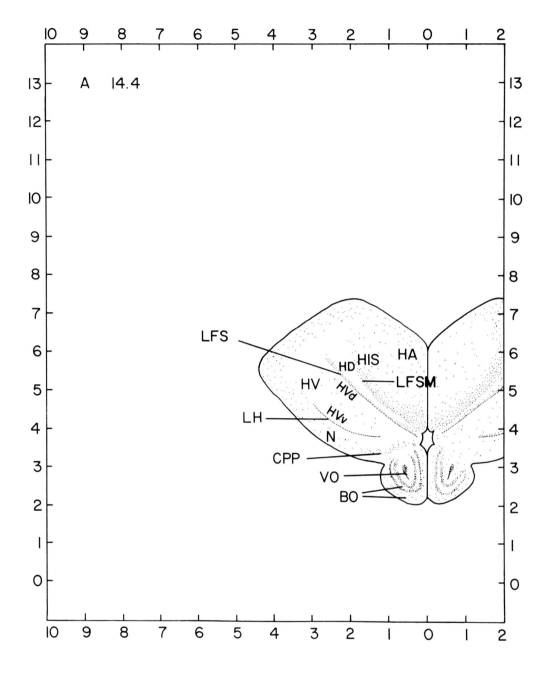

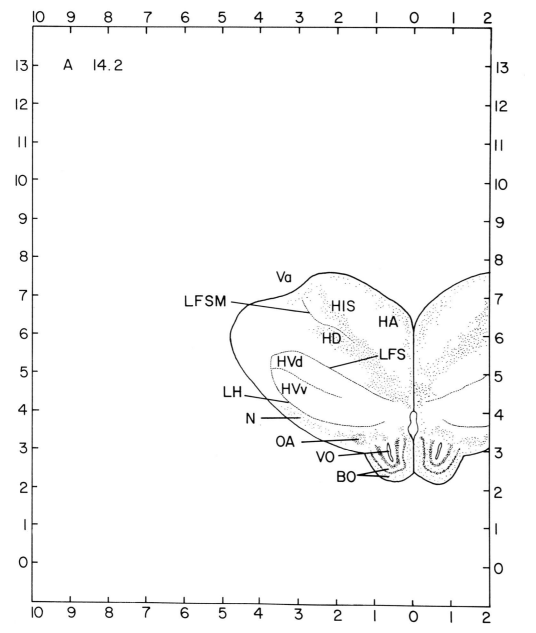

BO   Bulbus olfactorius
HA   Hyperstriatum accessorium
HD   Hyperstriatum dorsale
HIS   Hyperstriatum intercalatum supremum
HVd   Hyperstriatum ventrale, pars dorsalis
HVv   Hyperstriatum ventrale, pars ventralis
LFS   Lamina frontalis superior
LFSM Lamina frontalis suprema
LH   Lamina hyperstriatica
N   Neostriatum
OA   Nucleus olfactorius anterior
Va   Vallecula telencephali
VO   Ventriculus olfactorius

BO   Bulbus olfactorius
HA   Hyperstriatum accessorium
HD   Hyperstriatum dorsale
HIS   Hyperstriatum intercalatum supremum
HVd   Hyperstriatum ventrale, pars dorsalis
HVv   Hyperstriatum ventrale, pars ventralis
LFS   Lamina frontalis superior
LFSM   Lamina frontalis suprema
LH   Lamina hyperstriatica
N   Neostriatum
OA   Nucleus olfactorius anterior
Va   Vallecula telencephali
VO   Ventriculus olfactorius

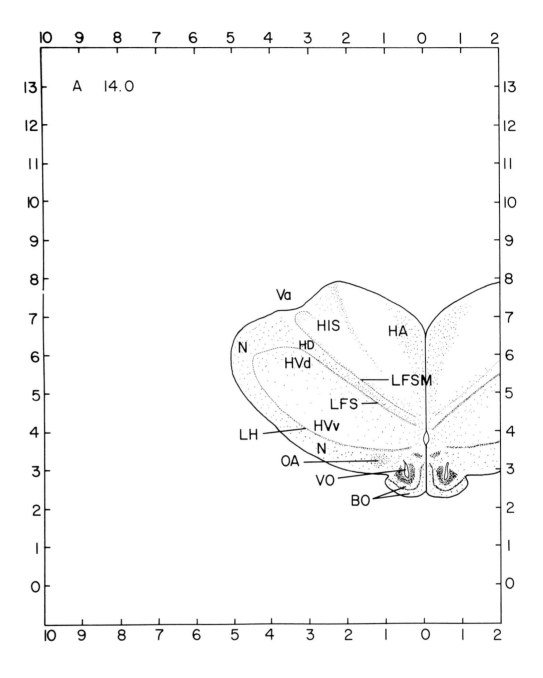

A   14.0

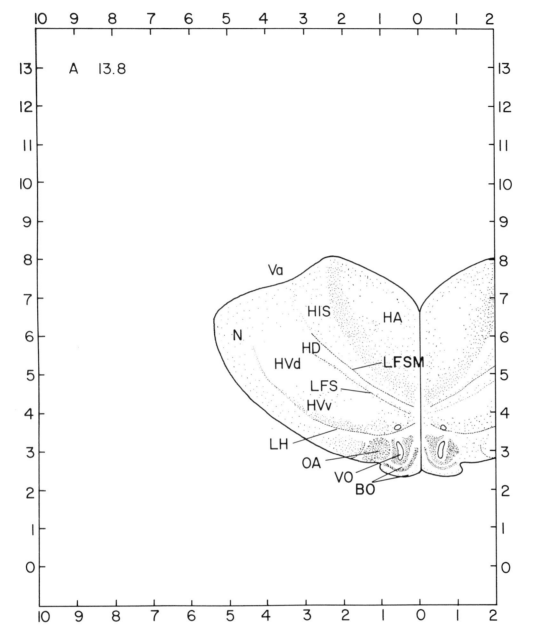

BO   Bulbus olfactorius
HA   Hyperstriatum accessorium
HD   Hyperstriatum dorsale
HIS   Hyperstriatum intercalatum supremum
HVd   Hyperstriatum ventrale, pars dorsalis
HVv   Hyperstriatum ventrale, pars ventralis
LFS   Lamina frontalis superior
LFSM Lamina frontalis suprema
LH   Lamina hyperstriatica
N   Neostriatum
OA   Nucleus olfactorius anterior
Va   Vallecula telencephali
VO   Ventriculus olfactorius

BO   Bulbus olfactorius
HA   Hyperstriatum accessorium
HD   Hyperstriatum dorsale
HIS   Hyperstriatum intercalatum supremum
HV   Hyperstriatum ventrale
HVd   Hyperstriatum ventrale, pars dorsalis
HVv   Hyperstriatum ventrale, pars ventralis
LFS   Lamina frontalis superior
LFSM   Lamina frontalis suprema
LH   Lamina hyperstriatica
N   Neostriatum
OA   Nucleus olfactorius anterior
Va   Vallecula telencephali
VO   Ventriculus olfactorius

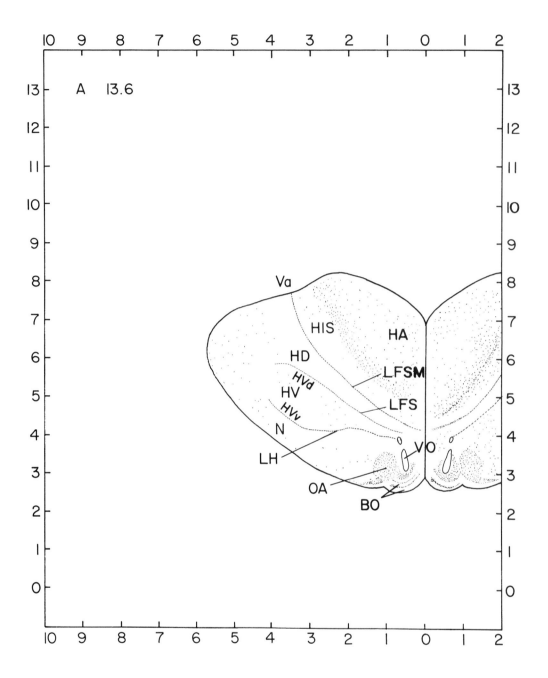

A   13.6

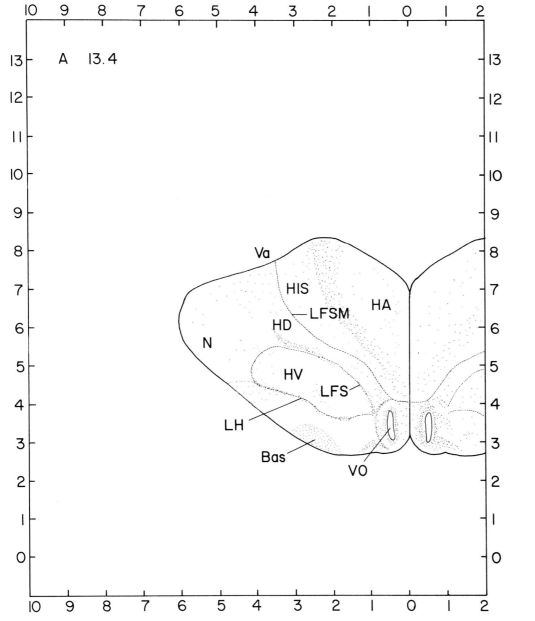

Bas   Nucleus basalis
HA   Hyperstriatum accessorium
HD   Hyperstriatum dorsale
HIS   Hyperstriatum intercalatum supremum
HV   Hyperstriatum ventrale
LFS   Lamina frontalis superior
LFSM Lamina frontalis suprema
LH   Lamina hyperstriatica
N   Neostriatum
Va   Vallecula telencephali
VO   Ventriculus olfactorius

Bas   Nucleus basalis
HA   Hyperstriatum accessorium
HD   Hyperstriatum dorsale
HIS   Hyperstriatum intercalatum supremum
HV   Hyperstriatum ventrale
LFS   Lamina frontalis superior
LFSM Lamina frontalis suprema
LH   Lamina hyperstriatica
N   Neostriatum
Va   Vallecula telencephali
VL   Ventriculus lateralis

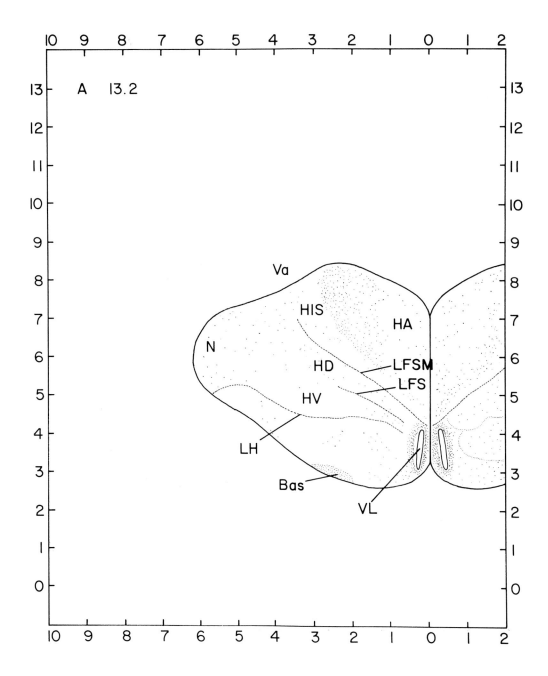

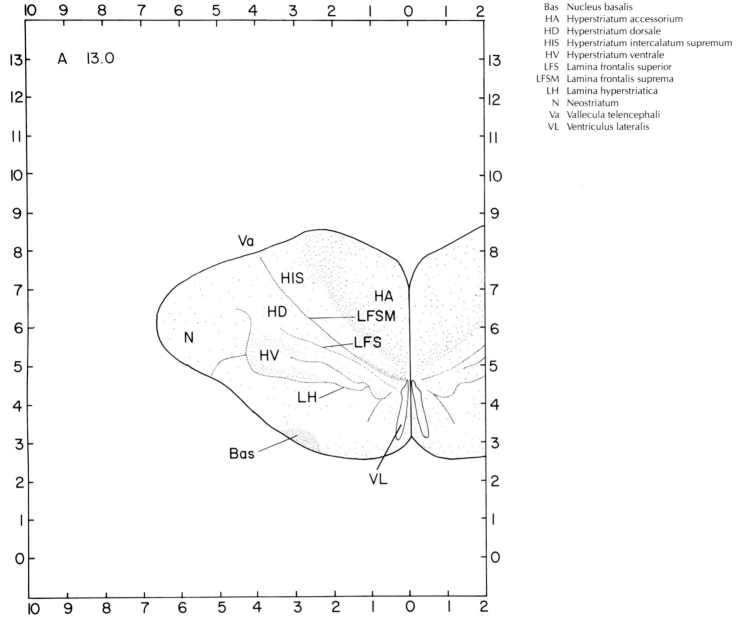

Bas   Nucleus basalis
HA   Hyperstriatum accessorium
HD   Hyperstriatum dorsale
HIS   Hyperstriatum intercalatum supremum
HV   Hyperstriatum ventrale
LFS   Lamina frontalis superior
LFSM   Lamina frontalis suprema
LH   Lamina hyperstriatica
N   Neostriatum
Va   Vallecula telencephali
VL   Ventriculus lateralis

Bas   Nucleus basalis
HA   Hyperstriatum accessorium
HD   Hyperstriatum dorsale
HIS   Hyperstriatum intercalatum supremum
HV   Hyperstriatum ventrale
LFS   Lamina frontalis superior
LFSM Lamina frontalis suprema
LH   Lamina hyperstriatica
N   Neostriatum
Va   Vallecula telencephali
VL   Ventriculus lateralis

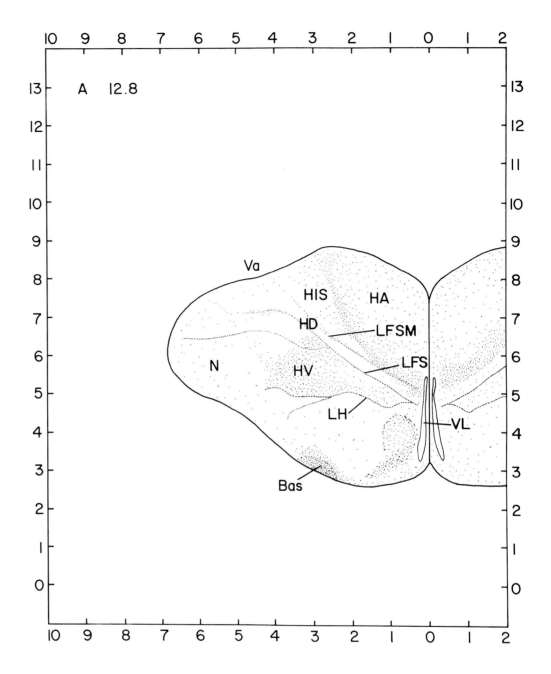

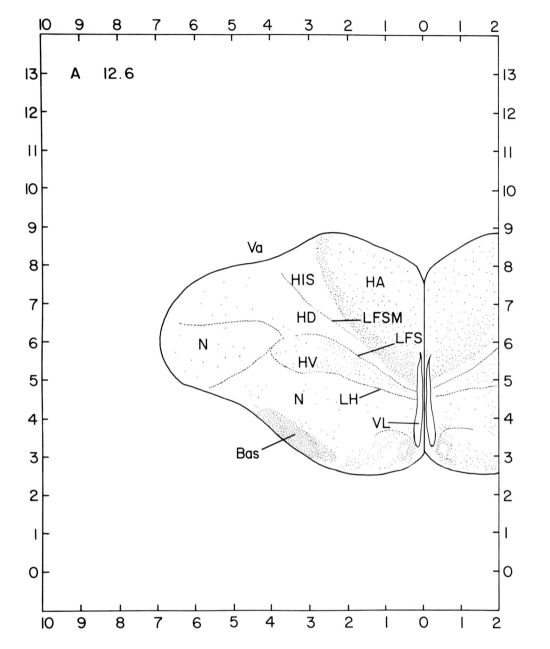

Bas   Nucleus basalis
HA   Hyperstriatum accessorium
HD   Hyperstriatum dorsale
HIS   Hyperstriatum intercalatum supremum
HV   Hyperstriatum ventrale
LFS   Lamina frontalis superior
LFSM   Lamina frontalis suprema
LH   Lamina hyperstriatica
N   Neostriatum
Va   Vallecula telencephali
VL   Ventriculus lateralis

Bas Nucleus basalis
HA Hyperstriatum accessorium
HD Hyperstriatum dorsale
HIS Hyperstriatum intercalatum supremum
HV Hyperstriatum ventrale
LFS Lamina frontalis superior
LFSM Lamina frontalis suprema
LH Lamina hyperstriatica
LMD Lamina medullaris dorsalis
LPO Lobus parolfactorius
N Neostriatum
Va Vallecula telencephali
VL Ventriculus lateralis

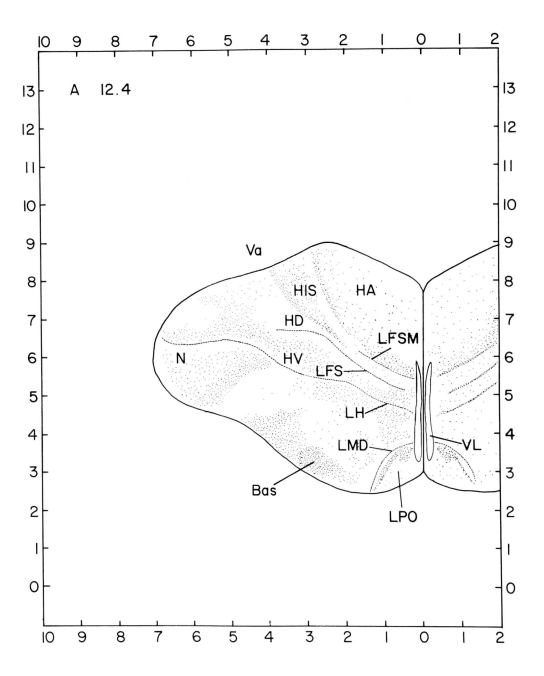

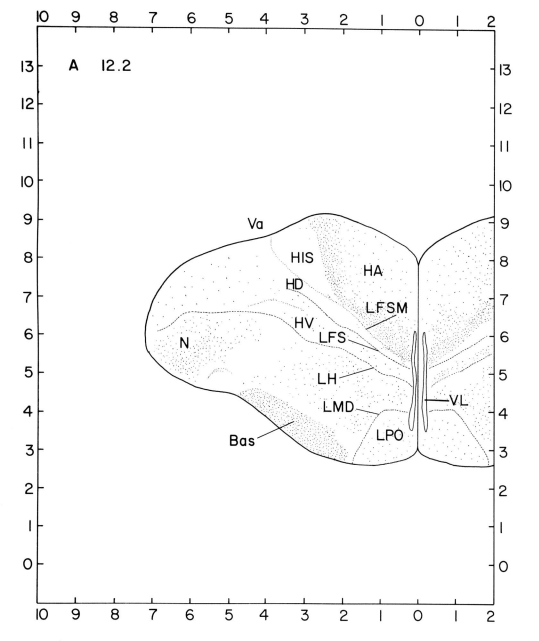

Bas   Nucleus basalis
HA   Hyperstriatum accessorium
HD   Hyperstriatum dorsale
HIS   Hyperstriatum intercalatum supremum
HV   Hyperstriatum ventrale
LFS   Lamina frontalis superior
LFSM   Lamina frontalis suprema
LH   Lamina hyperstriatica
LMD   Lamina medullaris dorsalis
LPO   Lobus parolfactorius
N   Neostriatum
Va   Vallecula telencephali
VL   Ventriculus lateralis

Bas   Nucleus basalis
HA   Hyperstriatum accessorium
HD   Hyperstriatum dorsale
HIS   Hyperstriatum intercalatum supremum
HV   Hyperstriatum ventrale
LFS   Lamina frontalis superior
LFSM   Lamina frontalis suprema
LH   Lamina hyperstriatica
LMD   Lamina medullaris dorsalis
LPO   Lobus parolfactorius
N   Neostriatum
Va   Vallecula telencephali
VL   Ventriculus lateralis

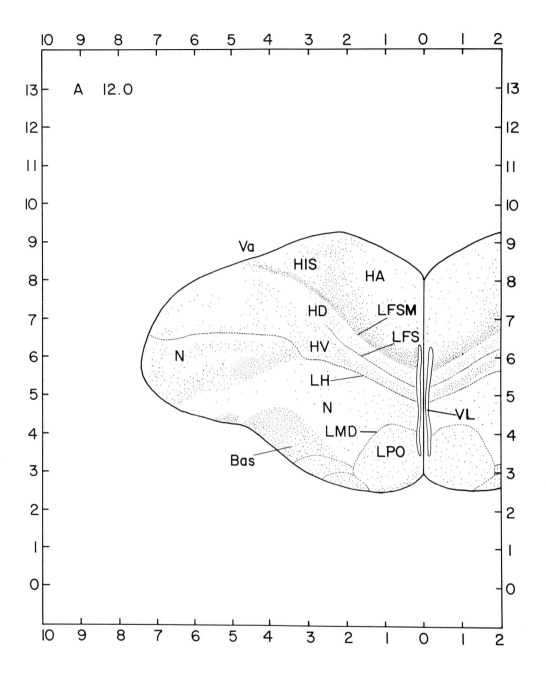

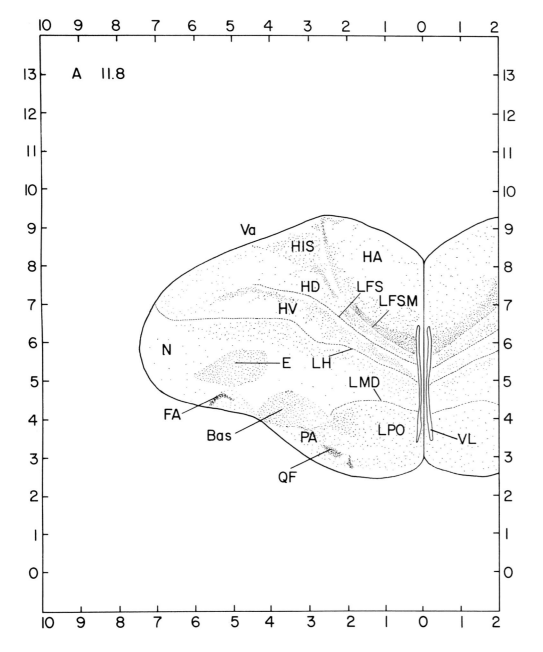

Bas   Nucleus basalis
  E   Ectostriatum
 FA   Tractus fronto-archistriaticus
 HA   Hyperstriatum accessorium
 HD   Hyperstriatum dorsale
HIS   Hyperstriatum intercalatum supremum
 HV   Hyperstriatum ventrale
LFS   Lamina frontalis superior
LFSM Lamina frontalis suprema
 LH   Lamina hyperstriatica
LMD   Lamina medullaris dorsalis
LPO   Lobus parolfactorius
  N   Neostriatum
 PA   Paleostriatum augmentatum (Caudate putamen)
 QF   Tractus quintofrontalis
 Va   Vallecula telencephali
 VL   Ventriculus lateralis

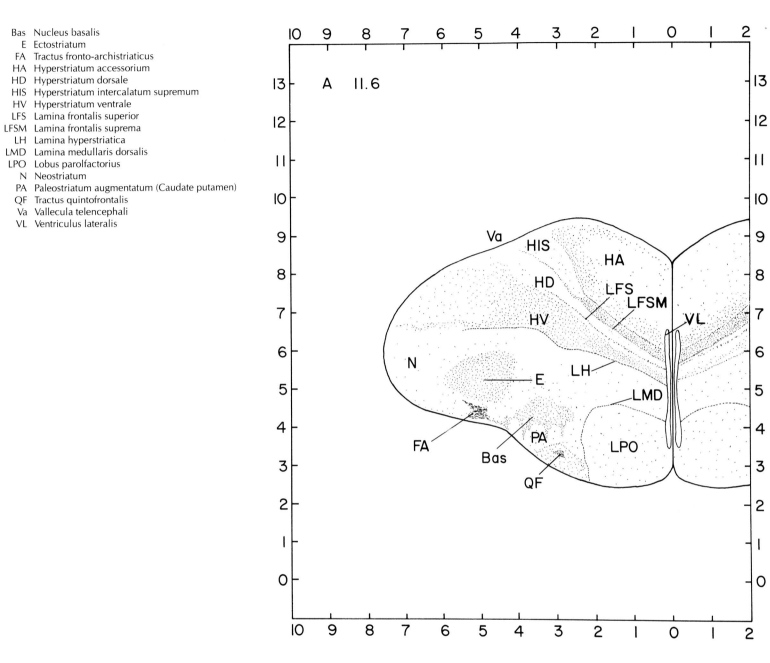

Bas　Nucleus basalis
E　Ectostriatum
FA　Tractus fronto-archistriaticus
HA　Hyperstriatum accessorium
HD　Hyperstriatum dorsale
HIS　Hyperstriatum intercalatum supremum
HV　Hyperstriatum ventrale
LFS　Lamina frontalis superior
LFSM　Lamina frontalis suprema
LH　Lamina hyperstriatica
LMD　Lamina medullaris dorsalis
LPO　Lobus parolfactorius
N　Neostriatum
PA　Paleostriatum augmentatum (Caudate putamen)
QF　Tractus quintofrontalis
Va　Vallecula telencephali
VL　Ventriculus lateralis

A 11.6

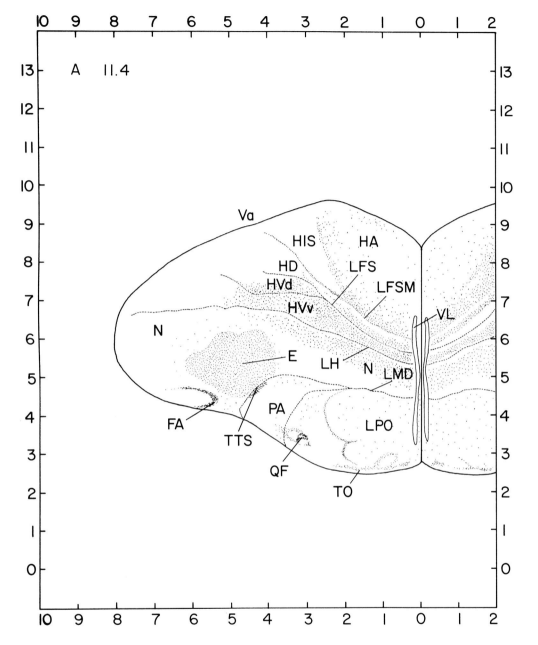

E   Ectostriatum
FA   Tractus fronto-archistriaticus
HA   Hyperstriatum accessorium
HD   Hyperstriatum dorsale
HIS   Hyperstriatum intercalatum supremum
HVd   Hyperstriatum ventrale, pars dorsalis
HVv   Hyperstriatum ventrale, pars ventralis
LFS   Lamina frontalis superior
LFSM   Lamina frontalis suprema
LH   Lamina hyperstriatica
LMD   Lamina medullaris dorsalis
LPO   Lobus parolfactorius
N   Neostriatum
PA   Paleostriatum augmentatum (Caudate putamen)
QF   Tractus quintofrontalis
TO   Tuberculum olfactorium
TTS   Tractus thalamostriaticus
Va   Vallecula telencephali
VL   Ventriculus lateralis

E   Ectostriatum
FA   Tractus fronto-archistriaticus
HA   Hyperstriatum accessorium
HD   Hyperstriatum dorsale
HIS Hyperstriatum intercalatum supremum
HV   Hyperstriatum ventrale
LFS Lamina frontalis superior
LFSM Lamina frontalis suprema
LH   Lamina hyperstriatica
LMD Lamina medullaris dorsalis
LPO Lobus parolfactorius
N   Neostriatum
PA   Paleostriatum augmentatum (Caudate putamen)
QF   Tractus quintofrontalis
TO   Tuberculum olfactorium
TTS Tractus thalamostriaticus
Va   Vallecula telencephali
VL   Ventriculus lateralis

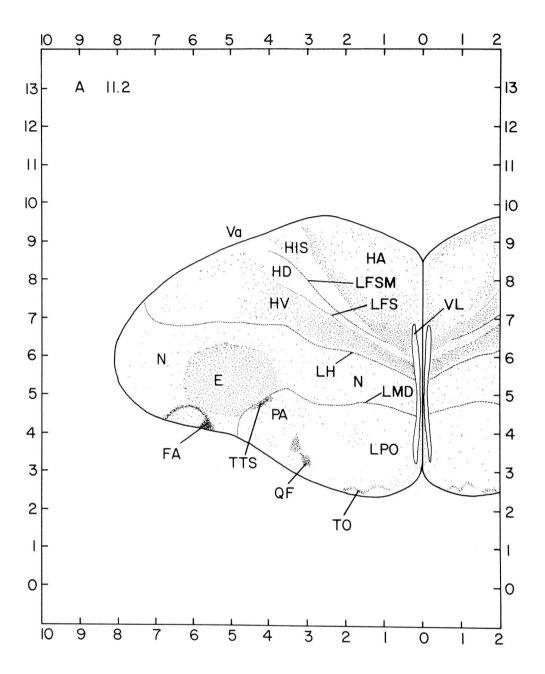

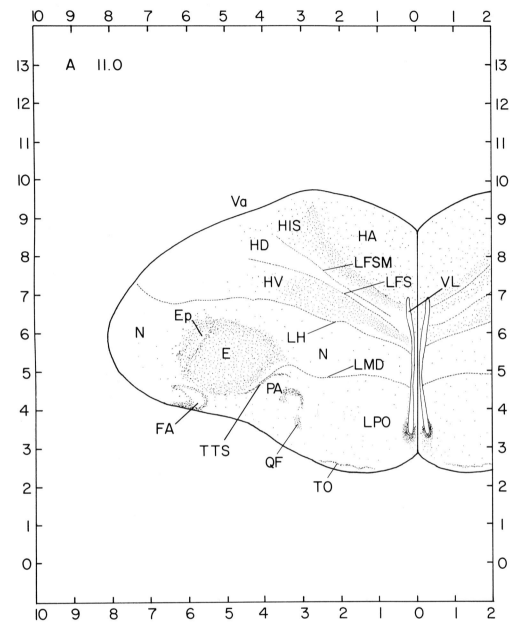

E   Ectostriatum
Ep   Cingulum periectostriatale (Periectostriatal belt)
FA   Tractus fronto-archistriaticus
HA   Hyperstriatum accessorium
HD   Hyperstriatum dorsale
HIS   Hyperstriatum intercalatum supremum
HV   Hyperstriatum ventrale
LFS   Lamina frontalis superior
LFSM Lamina frontalis suprema
LH   Lamina hyperstriatica
LMD   Lamina medullaris dorsalis
LPO   Lobus parolfactorius
N   Neostriatum
PA   Paleostriatum augmentatum (Caudate putamen)
QF   Tractus quintofrontalis
TO   Tuberculum olfactorium
TTS   Tractus thalamostriaticus
Va   Vallecula telencephali
VL   Ventriculus lateralis

E   Ectostriatum
Ep   Cingulum periectostriatale (Periectostriatal belt)
FA   Tractus fronto-archistriaticus
HA   Hyperstriatum accessorium
HD   Hyperstriatum dorsale
HIS Hyperstriatum intercalatum supremum
HV   Hyperstriatum ventrale
LFS Lamina frontalis superior
LFSM Lamina frontalis suprema
LH   Lamina hyperstriatica
LMD Lamina medullaris dorsalis
LPO Lobus parolfactorius
N   Neostriatum
PA   Paleostriatum augmentatum (Caudate putamen)
QF   Tractus quintofrontalis
TO   Tuberculum olfactorium
TTS Tractus thalamostriaticus
Va   Vallecula telencephali
VL   Ventriculus lateralis

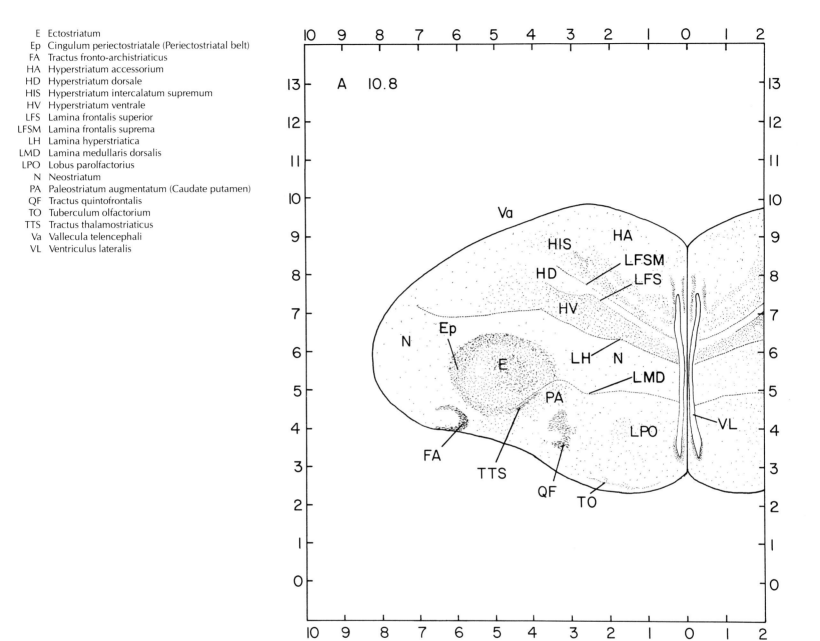

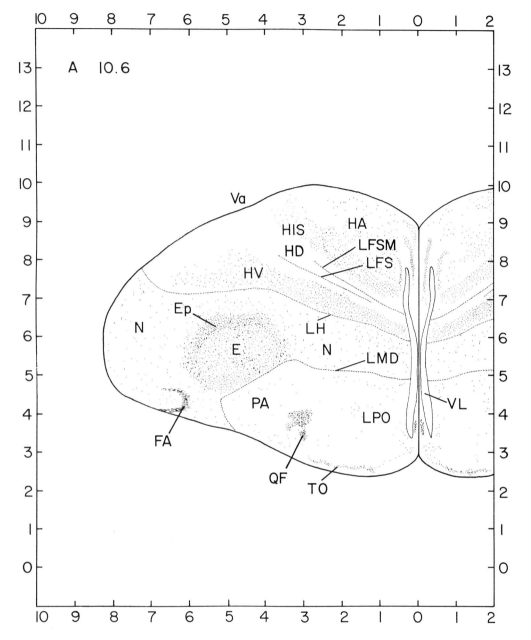

E   Ectostriatum
Ep   Cingulum periectostriatale (Periectostriatal belt)
FA   Tractus fronto-archistriaticus
HA   Hyperstriatum accessorium
HD   Hyperstriatum dorsale
HIS   Hyperstriatum intercalatum supremum
HV   Hyperstriatum ventrale
LFS   Lamina frontalis superior
LFSM   Lamina frontalis suprema
LH   Lamina hyperstriatica
LMD   Lamina medullaris dorsalis
LPO   Lobus parolfactorius
N   Neostriatum
PA   Paleostriatum augmentatum (Caudate putamen)
QF   Tractus quintofrontalis
TO   Tuberculum olfactorium
Va   Vallecula telencephali
VL   Ventriculus lateralis

E   Ectostriatum
Ep   Cingulum periectostriatale (Periectostriatal belt)
FA   Tractus fronto-archistriaticus
HA   Hyperstriatum accessorium
HD   Hyperstriatum dorsale
HIS   Hyperstriatum intercalatum supremum
HV   Hyperstriatum ventrale
LFS   Lamina frontalis superior
LFSM   Lamina frontalis suprema
LH   Lamina hyperstriatica
LMD   Lamina medullaris dorsalis
LPO   Lobus parolfactorius
N   Neostriatum
PA   Paleostriatum augmentatum (Caudate putamen)
QF   Tractus quintofrontalis
TO   Tuberculum olfactorium
Va   Vallecula telencephali
VL   Ventriculus lateralis

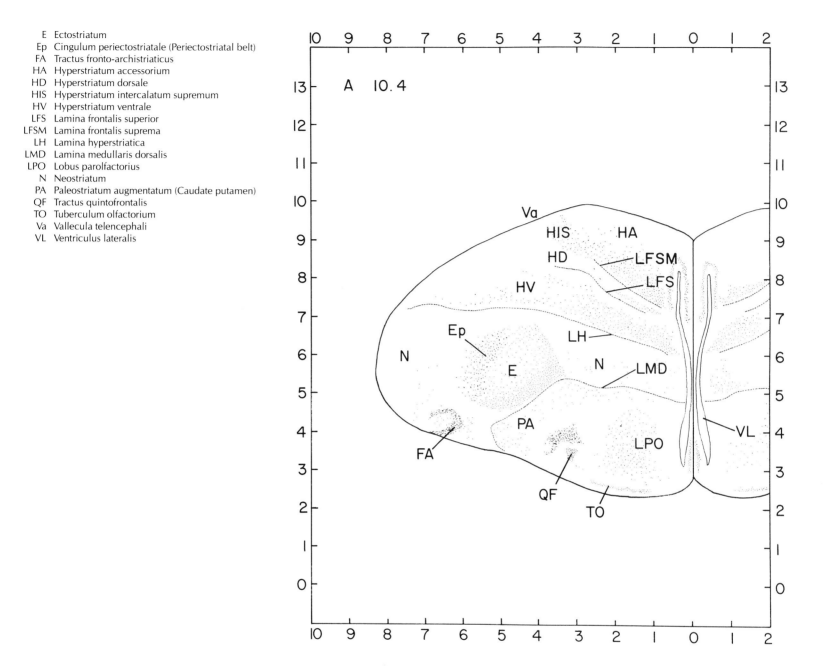

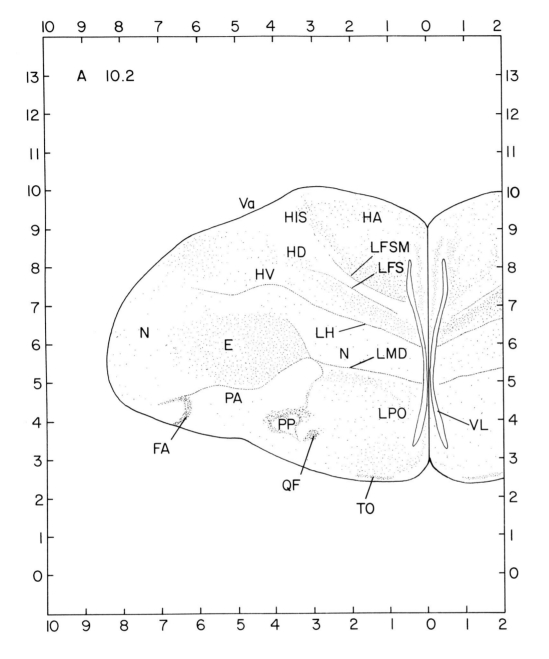

E   Ectostriatum
FA   Tractus fronto-archistriaticus
HA   Hyperstriatum accessorium
HD   Hyperstriatum dorsale
HIS Hyperstriatum intercalatum supremum
HV   Hyperstriatum ventrale
LFS Lamina frontalis superior
LFSM Lamina frontalis suprema
LH   Lamina hyperstriatica
LMD Lamina medullaris dorsalis
LPO Lobus parolfactorius
N   Neostriatum
PA   Paleostriatum augmentatum (Caudate putamen)
PP   Paleostriatum primitivum (Globus pallidus)
QF   Tractus quintofrontalis
TO   Tuberculum olfactorium
Va   Vallecula telencephali
VL   Ventriculus lateralis

Ac Nucleus accumbens
E Ectostriatum
FA Tractus fronto-archistriaticus
HA Hyperstriatum accessorium
HD Hyperstriatum dorsale
HV Hyperstriatum ventrale
LFS Lamina frontalis superior
LFSM Lamina frontalis suprema
LH Lamina hyperstriatica
LMD Lamina medullaris dorsalis
LPO Lobus parolfactorius
N Neostriatum
PA Paleostriatum augmentatum (Caudate putamen)
PP Paleostriatum primitivum (Globus pallidus)
QF Tractus quintofrontalis
TO Tuberculum olfactorium
Va Vallecula telencephali
VL Ventriculus lateralis

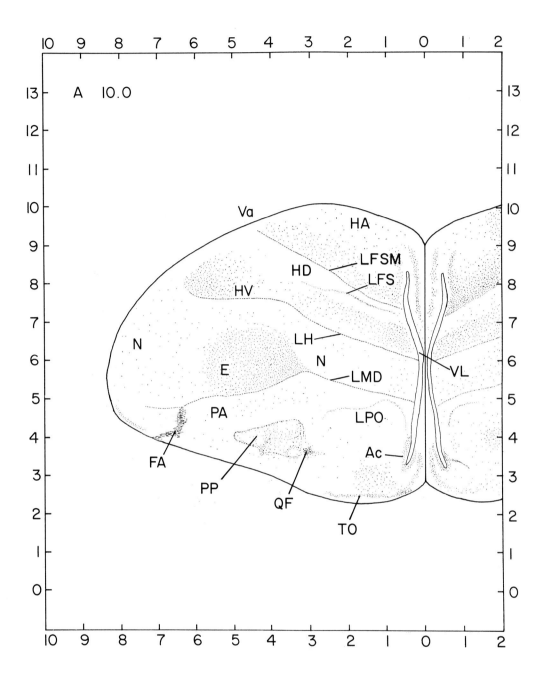

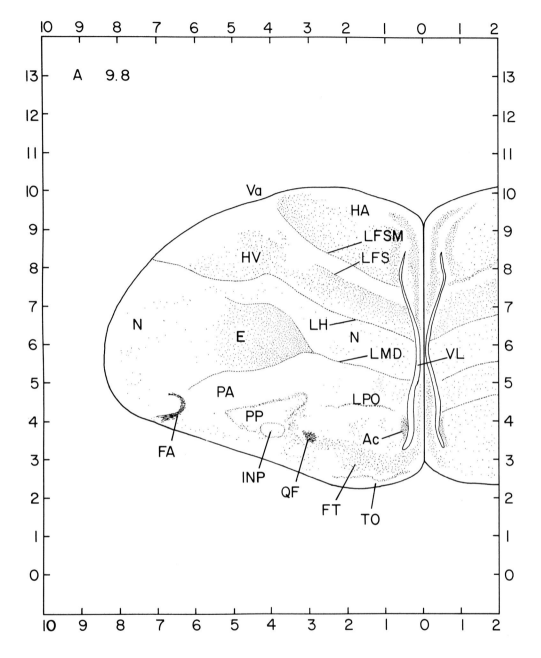

Ac Nucleus accumbens
 E Ectostriatum
FA Tractus fronto-archistriaticus
FT Tractus frontothalamicus et tractus
    thalamofrontalis
HA Hyperstriatum accessorium
HV Hyperstriatum ventrale
INP Nucleus intrapeduncularis
LFS Lamina frontalis superior
LFSM Lamina frontalis suprema
LH Lamina hyperstriatica
LMD Lamina medullaris dorsalis
LPO Lobus parolfactorius
 N Neostriatum
PA Paleostriatum augmentatum (Caudate putamen)
PP Paleostriatum primitivum (Globus pallidus)
QF Tractus quintofrontalis
TO Tuberculum olfactorium
Va Vallecula telencephali
VL Ventriculus lateralis

Ac   Nucleus accumbens
E   Ectostriatum
FA   Tractus fronto-archistriaticus
HA   Hyperstriatum accessorium
HV   Hyperstriatum ventrale
INP   Nucleus intrapeduncularis
LFS   Lamina frontalis superior
LFSM   Lamina frontalis suprema
LH   Lamina hyperstriatica
LMD   Lamina medullaris dorsalis
LPO   Lobus parolfactorius
LSO   Organum septi laterale (Lateral septal organ)
N   Neostriatum
PA   Paleostriatum augmentatum (Caudate putamen)
PP   Paleostriatum primitivum (Globus pallidus)
QF   Tractus quintofrontalis
TO   Tuberculum olfactorium
Va   Vallecula telencephali
VL   Ventriculus lateralis

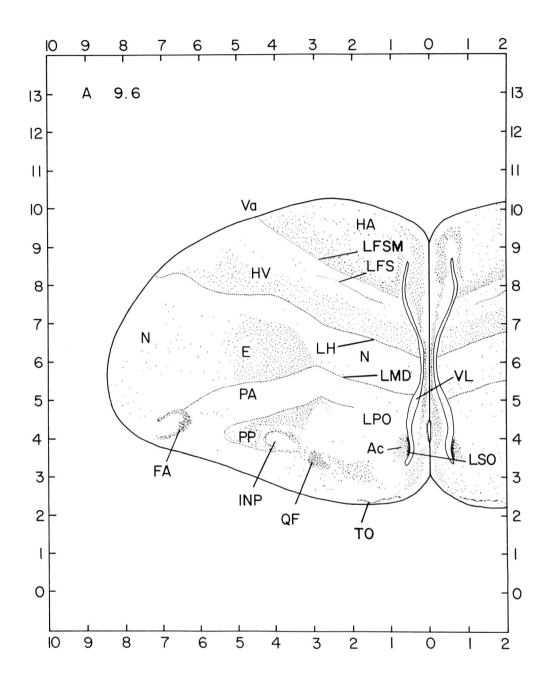

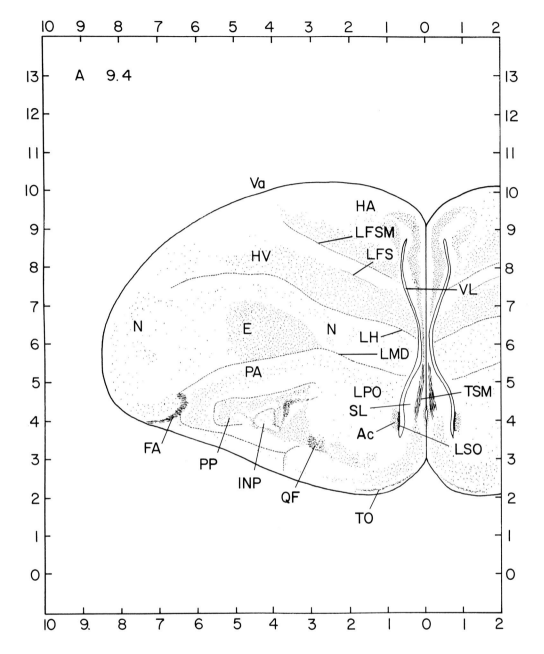

Ac   Nucleus accumbens
E   Ectostriatum
FA   Tractus fronto-archistriaticus
HA   Hyperstriatum accessorium
HV   Hyperstriatum ventrale
INP   Nucleus intrapeduncularis
LFS   Lamina frontalis superior
LFSM   Lamina frontalis suprema
LH   Lamina hyperstriatica
LMD   Lamina medullaris dorsalis
LPO   Lobus parolfactorius
LSO   Organum septi laterale (Lateral septal organ)
N   Neostriatum
PA   Paleostriatum augmentatum (Caudate putamen)
PP   Paleostriatum primitivum (Globus pallidus)
QF   Tractus quintofrontalis
SL   Nucleus septalis lateralis
TO   Tuberculum olfactorium
TSM   Tractus septomesencephalicus
Va   Vallecula telencephali
VL   Ventriculus lateralis

AA Archistriatum anterior [rostrale] (Zeier and
    Karten)
Ac Nucleus accumbens
E Ectostriatum
FA Tractus fronto-archistriaticus
FDB Fasciculus diagonalis Brocae
FPL Fasciculus prosencephali lateralis (Lateral
    forebrain bundle)
HA Hyperstriatum accessorium
HV Hyperstriatum ventrale
INP Nucleus intrapeduncularis
LFS Lamina frontalis superior
LFSM Lamina frontalis suprema
LH Lamina hyperstriatica
LMD Lamina medullaris dorsalis
LPO Lobus parolfactorius
LSO Organum septi laterale (Lateral septal organ)
N Neostriatum
NI Neostriatum intermedium
PA Paleostriatum augmentatum (Caudate putamen)
POD Nucleus preopticus dorsolateralis
PP Paleostriatum primitivum (Globus pallidus)
PVT Paleostriatum ventrale (Kitt and Brauth)
QF Tractus quintofrontalis
SL Nucleus septalis lateralis
SOv Nucleus supraopticus (Ralph), pars ventralis
TSM Tractus septomesencephalicus
VL Ventriculus lateralis

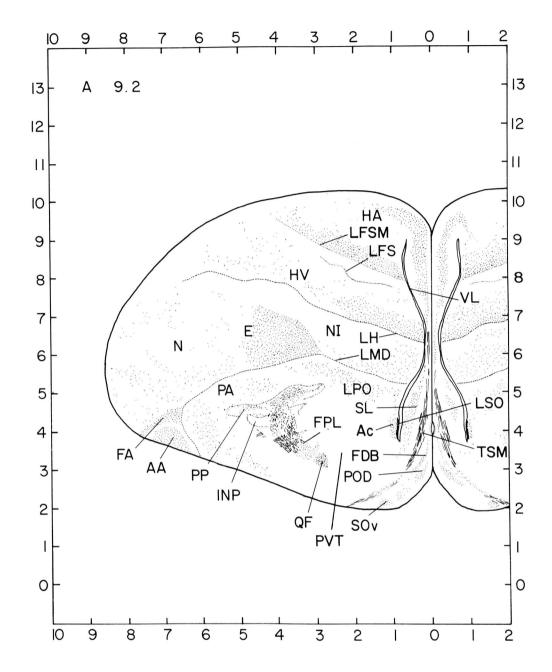

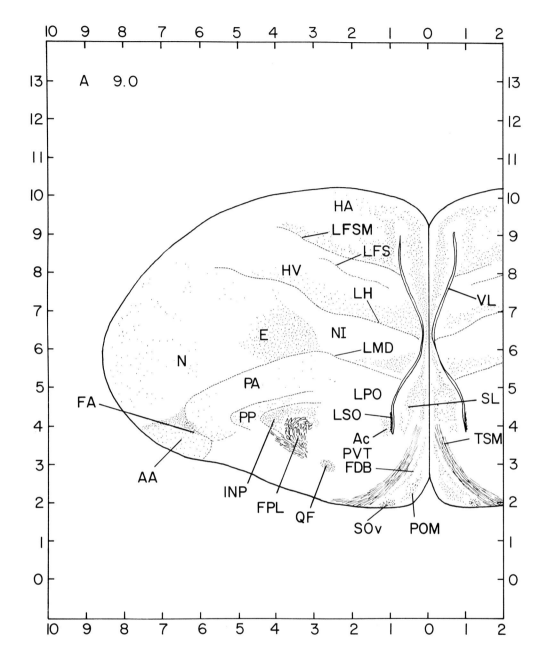

AA   Archistriatum anterior [rostrale] (Zeier and
      Karten)
Ac   Nucleus accumbens
E   Ectostriatum
FA   Tractus fronto-archistriaticus
FDB Fasciculus diagonalis Brocae
FPL Fasciculus prosencephali lateralis (Lateral
      forebrain bundle)
HA   Hyperstriatum accessorium
HV   Hyperstriatum ventrale
INP Nucleus intrapeduncularis
LFS Lamina frontalis superior
LFSM Lamina frontalis suprema
LH   Lamina hyperstriatica
LMD Lamina medullaris dorsalis
LPO Lobus parolfactorius
LSO Organum septi laterale (Lateral septal organ)
N   Neostriatum
NI   Neostriatum intermedium
PA   Paleostriatum augmentatum (Caudate putamen)
POM Nucleus preopticus medialis (van Tienhoven)
PP   Paleostriatum primitivum (Globus pallidus)
PVT Paleostriatum ventrale (Kitt and Brauth)
QF   Tractus quintofrontalis
SL   Nucleus septalis lateralis
SOv Nucleus supraopticus (Ralph), pars ventralis
TSM Tractus septomesencephalicus
VL   Ventriculus lateralis

AA Archistriatum anterior [rostrale] (Zeier and
    Karten)
 Ac Nucleus accumbens
  E Ectostriatum
 FA Tractus fronto-archistriaticus
FDB Fasciculus diagonalis Brocae
FPL Fasciculus prosencephali lateralis (Lateral
    forebrain bundle)
 HA Hyperstriatum accessorium
 HV Hyperstriatum ventrale
INP Nucleus intrapeduncularis
LFS Lamina frontalis superior
LFSM Lamina frontalis suprema
 LH Lamina hyperstriatica
LMD Lamina medullaris dorsalis
LSO Organum septi laterale (Lateral septal organ)
MPO Nucleus magnocellularis preopticus (van
    Tienhoven)
 (d) pars dorsalis
 (m) pars medialis
 (v) pars ventralis
  N Neostriatum
 NI Neostriatum intermedium
 PA Paleostriatum augmentatum (Caudate putamen)
POM Nucleus preopticus medialis (van Tienhoven)
POP Nucleus preopticus periventricularis
 PP Paleostriatum primitivum (Globus pallidus)
PVT Paleostriatum ventrale (Kitt and Brauth)
 QF Tractus quintofrontalis
 SL Nucleus septalis lateralis
SOe Nucleus supraopticus (Ralph), pars externus
SOv Nucleus supraopticus (Ralph), pars ventralis
TSM Tractus septomesencephalicus
 VL Ventriculus lateralis

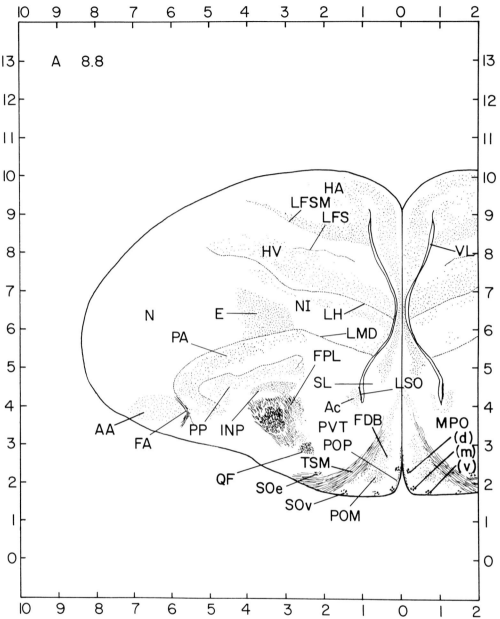

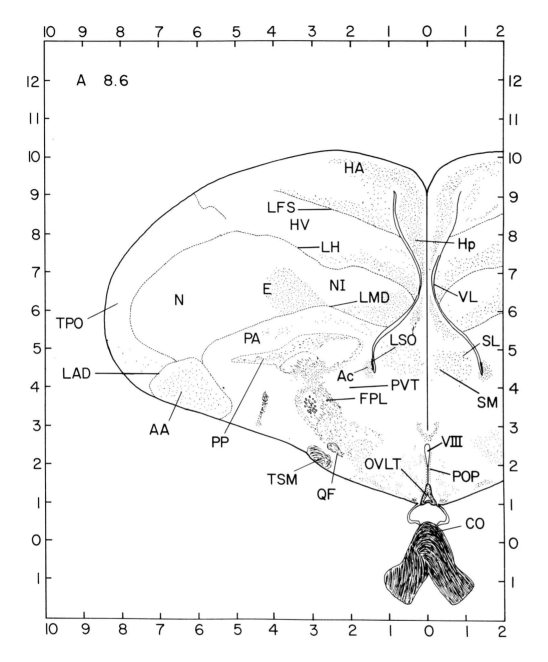

A 8.6

AA	Archistriatum anterior [rostrale] (Zeier and Karten)
Ac	Nucleus accumbens
CO	Chiasma opticum
E	Ectostriatum
FPL	Fasciculus prosencephali lateralis (Lateral forebrain bundle)
HA	Hyperstriatum accessorium
Hp	Hippocampus
HV	Hyperstriatum ventrale
LAD	Lamina archistriatalis dorsalis
LFS	Lamina frontalis superior
LH	Lamina hyperstriatica
LMD	Lamina medullaris dorsalis
LSO	Organum septi laterale (Lateral septal organ)
N	Neostriatum
NI	Neostriatum intermedium
OVLT	Organum vasculosum lamina terminalis
PA	Paleostriatum augmentatum (Caudate putamen)
POP	Nucleus preopticus periventricularis
PP	Paleostriatum primitivum (Globus pallidus)
PVT	Paleostriatum ventrale (Kitt and Brauth)
QF	Tractus quintofrontalis
SL	Nucleus septalis lateralis
SM	Nucleus septalis medialis
TPO	Area temporo-parieto-occipitalis (Edinger, Wallenberg, and Holmes)
TSM	Tractus septomesencephalicus
VL	Ventriculus lateralis
V III	Ventriculus tertius (Third ventricle)

AA   Archistriatum anterior [rostrale] (Zeier and Karten)
Ac   Nucleus accumbens
AM   Nucleus anterior [rostralis] medialis hypothalami
CA   Commissura anterior [rostralis] (Anterior commissure)
CO   Chiasma opticum
CPi Cortex piriformis
E   Ectostriatum
FPL Fasciculus prosencephali lateralis (Lateral forebrain bundle)
HA   Hyperstriatum accessorium
Hp   Hippocampus
HV   Hyperstriatum ventrale
LAD Lamina archistriatalis dorsalis
LFS Lamina frontalis superior
LH   Lamina hyperstriatica
LMD Lamina medullaris dorsalis
LSO Organum septi laterale (Lateral septal organ)
N   Neostriatum
nCPa Nucleus commissurae pallii (Bed nucleus pallial commissure)
NI   Neostriatum intermedium
PA   Paleostriatum augmentatum (Caudate putamen)
POP Nucleus preopticus periventricularis
PP   Paleostriatum primitivum (Globus pallidus)
PVT Paleostriatum ventrale (Kitt and Brauth)
QF   Tractus quintofrontalis
SCNm Nucleus suprachiasmaticus, pars medialis
SL   Nucleus septalis lateralis
SM   Nucleus septalis medialis
TPO Area temporo-parieto-occipitalis (Edinger, Wallenberg, and Holmes)
TSM Tractus septomesencephalicus
VL   Ventriculus lateralis
VLT Nucleus ventrolateralis thalami
V III Ventriculus tertius (Third ventricle)

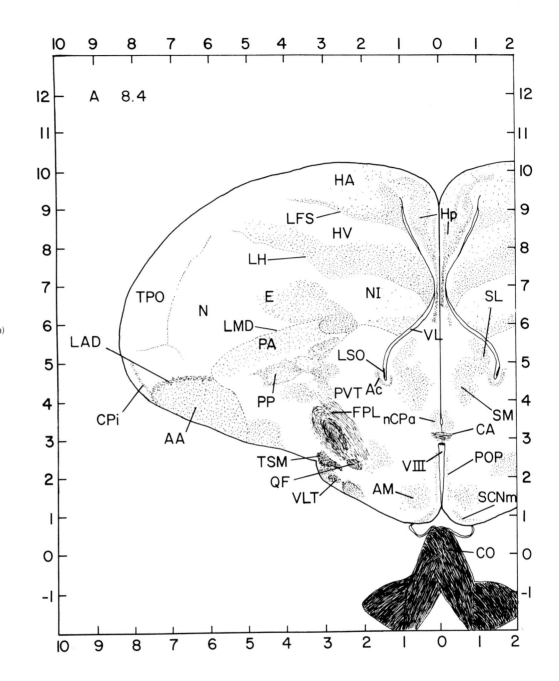

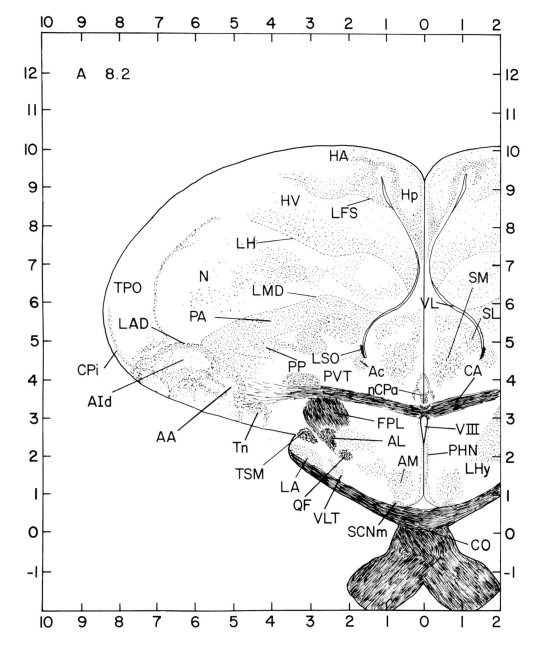

AA   Archistriatum anterior [rostrale] (Zeier and Karten)
Ac   Nucleus accumbens
AId Archistriatum intermedium, pars dorsalis (Zeier and Karten)
AL   Ansa lenticularis
AM   Nucleus anterior [rostralis] medialis hypothalami
CA   Commissura anterior [rostralis] (Anterior commissure)
CO   Chiasma opticum
CPi Cortex piriformis
FPL Fasciculus prosencephali lateralis (Lateral forebrain bundle)
HA   Hyperstriatum accessorium
Hp   Hippocampus
HV   Hyperstriatum ventrale
LA   Nucleus lateralis anterior [rostralis] thalami
LAD Lamina archistriatalis dorsalis
LFS Lamina frontalis superior
LH   Lamina hyperstriatica
LHy Regio lateralis hypothalami (Lateral hypothalamic area)
LMD Lamina medullaris dorsalis
LSO Organum septi laterale (Lateral septal organ)
N   Neostriatum
nCPa Nucleus commissurae pallii (Bed nucleus pallial commissure)
PA   Paleostriatum augmentatum (Caudate putamen)
PHN Nucleus periventricularis hypothalami
PP   Paleostriatum primitivum (Globus pallidus)
PVT Paleostriatum ventrale (Kitt and Brauth)
QF   Tractus quintofrontalis
SCNm Nucleus suprachiasmaticus, pars medialis
SL   Nucleus septalis lateralis
SM   Nucleus septalis medialis
Tn   Nucleus taeniae
TPO Area temporo-parieto-occipitalis (Edinger, Wallenberg, and Holmes)
TSM Tractus septomesencephalicus
VL   Ventriculus lateralis
VLT Nucleus ventrolateralis thalami
V III Ventriculus tertius (Third ventricle)

AId Archistriatum intermedium, pars dorsalis
       (Zeier and Karten)
AIv Archistriatum intermedium, pars ventralis
       (Zeier and Karten)
AL Ansa lenticularis
AM Nucleus anterior [rostralis] medialis
       hypothalami
CO Chiasma opticum
CPa Commissura pallii
CPi Cortex piriformis
DLAmc Nucleus dorsolateralis anterior [rostralis]
       thalami, pars magnocellularis
DSv Nucleus decussationis supraopticae, pars
       ventralis
F Fornix
FPL Fasciculus prosencephali lateralis (Lateral
       forebrain bundle)
HA Hyperstriatum accessorium
Hp Hippocampus
HV Hyperstriatum ventrale
LA Nucleus lateralis anterior thalami
LAD Lamina archistriatalis dorsalis
LFS Lamina frontalis superior
LH Lamina hyperstriatica
LHy Regio lateralis hypothalami (Lateral
       hypothalamic area)
LMD Lamina medullaris dorsalis
LSO Organum septi laterale (Lateral septal organ)
N Neostriatum
nCPa Nucleus commissurae pallii (Bed nucleus
       pallial commissure)
OM Tractus occipitomesencephalicus
PA Paleostriatum augmentatum (Caudate
       putamen)
PHN Nucleus periventricularis hypothalami
PP Paleostriatum primitivum (Globus pallidus)
PVN Nucleus paraventricularis magnocellularis
       (Paraventricular nucleus)
PVT Paleostriatum ventrale (Kitt and Brauth)
QF Tractus quintofrontalis
RSd Nucleus reticularis superior, pars dorsalis
SL Nucleus septalis lateralis
SM Nucleus septalis medialis
SSO Organum subseptale (Subseptal organ [Legait
       and Legait]); organum interventriculare
       (Interventricular organ [Blähser])
Tn Nucleus taeniae
TPO Area temporo-parieto-occipitalis (Edinger,
       Wallenberg, and Holmes)
TSM Tractus septomesencephalicus
VL Ventriculus lateralis
V III Ventriculus tertius (Third ventricle)
VLT Nucleus ventrolateralis thalami

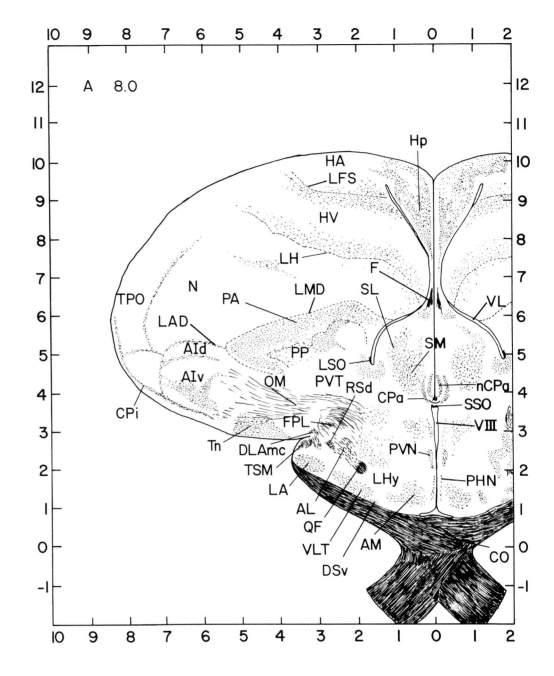

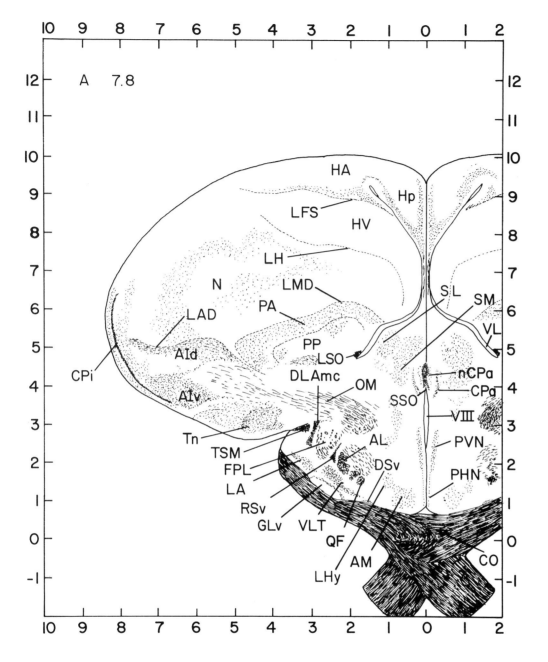

AId Archistriatum intermedium, pars dorsalis (Zeier and Karten)
AIv Archistriatum intermedium, pars ventralis (Zeier and Karten)
AL Ansa lenticularis
AM Nucleus anterior medialis [rostralis] hypothalami
CO Chiasma opticum
CPa Commissura pallii
CPi Cortex piriformis
DLAmc Nucleus dorsolateralis anterior [rostralis] thalami, pars magnocellularis
DSv Nucleus decussationis supraopticae, pars ventralis
FPL Fasciculus prosencephali lateralis (Lateral forebrain organ)
GLv Nucleus geniculatus lateralis, pars ventralis
HA Hyperstriatum accessorium
Hp Hippocampus
HV Hyperstriatum ventrale
LA Nucleus lateralis anterior thalami
LAD Lamina archistriatalis dorsalis
LFS Lamina frontalis superior
LH Lamina hyperstriatica
LHy Regio lateralis hypothalami (Lateral hypothalamic area)
LMD Lamina medullaris dorsalis
LSO Organum septi laterale (Lateral septal organ)
N Neostriatum
nCPa Nucleus commissurae pallii (Bed nucleus pallial commissure)
OM Tractus occipitomesencephalicus
PA Paleostriatum augmentatum (Caudate putamen)
PHN Nucleus periventricularis hypothalami
PP Paleostriatum primitivum (Globus pallidus)
PVN Nucleus paraventricularis magnocellularis (Paraventricular nucleus)
QF Tractus quintofrontalis
RSv Nucleus reticularis superior, pars ventralis
SL Nucleus septalis lateralis
SM Nucleus septalis medialis
SSO Organum subseptale (Subseptal organ [Legait and Legait]); organum interventriculare (Interventricular organ [Blähser])
Tn Nucleus taeniae
TSM Tractus septomesencephalicus
VL Ventriculus lateralis
V III Ventriculus tertius (Third ventricle)
VLT Nucleus ventrolateralis thalami

AId   Archistriatum intermedium, pars dorsalis (Zeier and Karten)
AIv   Archistriatum intermedium, pars ventralis (Zeier and Karten)
AL   Ansa lenticularis
Am   Archistriatum mediale (Zeier and Karten)
AM   Nucleus anterior medialis [rostralis] hypothalami
CO   Chiasma opticum
CPi   Cortex piriformis
DLAmc   Nucleus dorsolateralis anterior [rostralis] thalami, pars magnocellularis
FPL   Fasciculus prosencephali lateralis (Lateral forebrain bundle)
GLv   Nucleus geniculatus lateralis, pars ventralis
HA   Hyperstriatum accessorium
Hp   Hippocampus
HV   Hyperstriatum ventrale
LA   Nucleus lateralis anterior thalami
LAD   Lamina archistriatalis dorsalis
LFS   Lamina frontalis superior
LH   Lamina hyperstriatica
LHy   Regio lateralis hypothalami (Lateral hypothalamic area)
LMD   Lamina medullaris dorsalis
N   Neostriatum
nST   Nucleus striae terminalis (Bed nucleus, stria terminalis)
OM   Tractus occipitomesencephalicus
PA   Paleostriatum augmentatum (Caudate putamen)
PCVL   Plexus choroideus ventriculi lateralis (Choroid plexus within lateral ventricle)
PHN   Nucleus periventricularis hypothalami
PP   Paleostriatum primitivum (Globus pallidus)
PVN   Nucleus paraventricularis magnocellularis (Paraventricular nucleus)
QF   Tractus quintofrontalis
RSd   Nucleus reticularis superior, pars dorsalis
RSv   Nucleus reticularis superior, pars ventralis
SL   Nucleus septalis lateralis
SM   Nucleus septalis medialis
SSO   Organum subseptale (Subseptal organ [Legait and Legait]); organum interventriculare (Interventricular organ [Blähser])
Tn   Nucleus taeniae
TSM   Tractus septomesencephalicus
VL   Ventriculus lateralis
V III   Ventriculus tertius (Third ventricle)

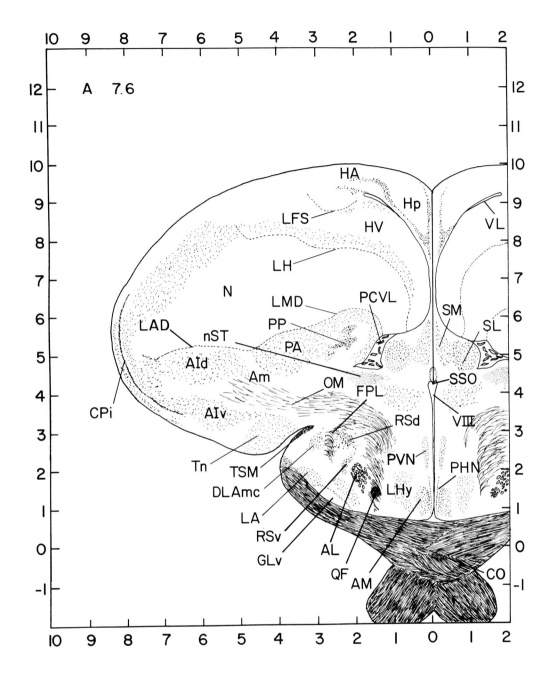

A 7.6

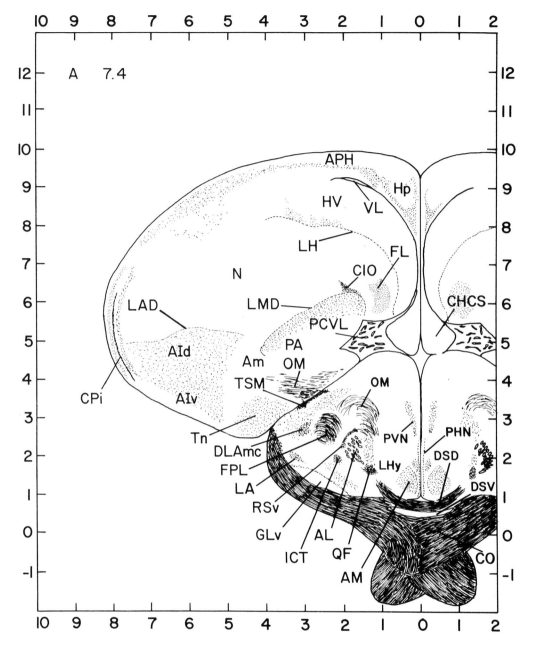

AId   Archistriatum intermedium, pars dorsalis
        (Zeier and Karten)
AIv   Archistriatum intermedium, pars ventralis
        (Zeier and Karten)
AL   Ansa lenticularis
Am   Archistriatum mediale (Zeier and Karten)
AM   Nucleus anterior medialis hypothalami
APH   Area parahippocampalis
CHCS Tractus cortico-habenularis et cortico-septalis
CIO   Capsula interna occipitalis
CO   Chiasma opticum
CPi   Cortex piriformis
DLAmc Nucleus dorsolateralis anterior [rostralis]
        thalami, pars magnocellularis
DSD   Decussatio supraoptica dorsalis
DSV   Decussatio supraoptica ventralis
FL   Field L
FPL   Fasciculus prosencephali lateralis (Lateral
        forebrain bundle)
GLv   Nucleus geniculatus lateralis, pars ventralis
Hp   Hippocampus
HV   Hyperstriatum ventrale
ICT   Nucleus intercalatus thalami
LA   Nucleus lateralis anterior thalami
LAD   Lamina archistriatalis dorsalis
LH   Lamina hyperstriatica
LHy   Regio lateralis hypothalami (Lateral
        hypothalamic area)
LMD   Lamina medullaris dorsalis
N   Neostriatum
OM   Tractus occipitomesencephalicus
PA   Paleostriatum augmentatum (Caudate
        putamen)
PCVL Plexus choroideus ventriculi lateralis (Choroid
        plexus within lateral ventricle)
PHN   Nucleus periventricularis hypothalami
PVN   Nucleus paraventricularis magnocellularis
        (Paraventricular nucleus)
QF   Tractus quintofrontalis
RSv   Nucleus reticularis superior, pars ventralis
Tn   Nucleus taeniae
TSM   Tractus septomesencephalicus
VL   Ventriculus lateralis

Ald   Archistriatum intermedium, pars dorsalis (Zeier
        and Karten)
Alv   Archistriatum intermedium, pars ventralis
        (Zeier and Karten)
AL   Ansa lenticularis
Am   Archistriatum mediale (Zeier and Karten)
APH   Area parahippocampalis
CDL   Area corticoidea dorsolateralis
CHCS Tractus cortico-habenularis et cortico-septalis
CIO   Capsula interna occipitalis
CO   Chiasma opticum
CPi   Cortex piriformis
DA   Tractus dorso-archistriaticus
DLAl Nucleus dorsolateralis anterior [rostralis]
        thalami, pars lateralis
DLAm Nucleus dorsolateralis anterior [rostralis]
        thalami, pars medialis
DSD   Decussatio supraoptica dorsalis
DSV   Decussatio supraoptica ventralis
FL   Field L
FPL   Fasciculus prosencephali lateralis (Lateral
        forebrain bundle)
GLv   Nucleus geniculatus lateralis, pars ventralis
Hp   Hippocampus
HV   Hyperstriatum ventrale
ICT   Nucleus intercalatus thalami
LAD   Lamina archistriatalis dorsalis
LH   Lamina hyperstriatica
LHy   Regio lateralis hypothalami (Lateral
        hypothalamic area)
LMD   Lamina medullaris dorsalis
N   Neostriatum
OM   Tractus occipitomesencephalicus
PA   Paleostriatum augmentatum (Caudate putamen)
PCVL Plexus choroideus ventriculi lateralis (Choroid
        plexus within lateral ventricle)
PHN   Nucleus periventricularis hypothalami
PVN   Nucleus paraventricularis magnocellularis
        (Paraventricular nucleus)
QF   Tractus quintofrontalis
ROT   Nucleus rotundus
RSd   Nucleus reticularis superior, pars dorsalis
Tn   Nucleus taeniae
TrO   Tractus opticus
TSM   Tractus septomesencephalicus
VL   Ventriculus lateralis
VMN   Nucleus ventromedialis hypothalami

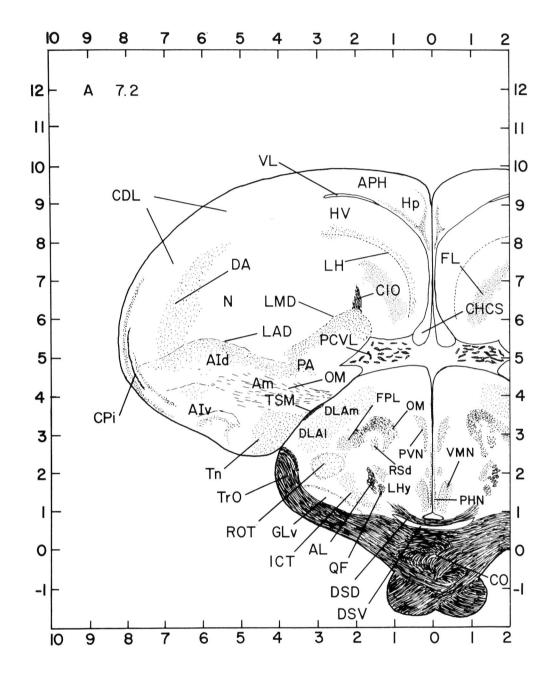

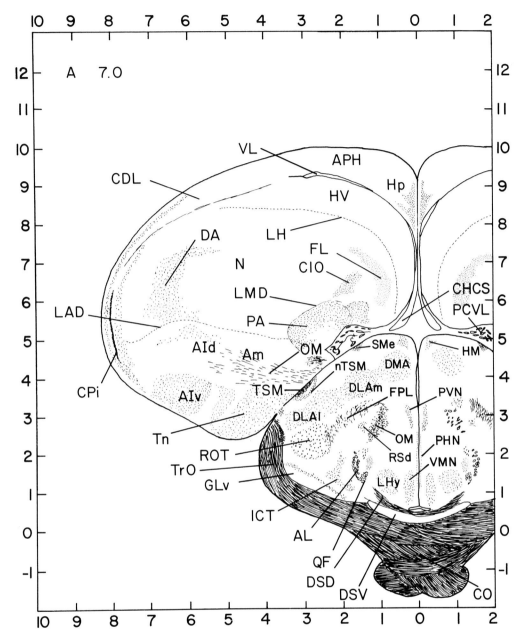

AId   Archistriatum intermedium, pars dorsalis (Zeier and Karten)
AIv   Archistriatum intermedium, pars ventralis (Zeier and Karten)
AL   Ansa lenticularis
Am   Archistriatum mediale (Zeier and Karten)
APH   Area parahippocampalis
CDL   Area corticoidea dorsolateralis
CHCS Tractus cortico-habenularis et cortico-septalis
CIO   Capsula interna occipitalis
CO   Chiasma opticum
CPi   Cortex piriformis
DA   Tractus dorso-archistriaticus
DLAl Nucleus dorsolateralis anterior [rostralis] thalami, pars lateralis
DLAm Nucleus dorsolateralis anterior [rostralis] thalami, pars medialis
DMA   Nucleus dorsomedialis anterior [rostralis] thalami
DSD   Decussatio supraoptica dorsalis
DSV   Decussatio supraoptica ventralis
FL   Field L
FPL   Fasciculus prosencephali lateralis (Lateral forebrain bundle)
GLv   Nucleus geniculatus lateralis, pars ventralis
HM   Nucleus habenularis medialis
Hp   Hippocampus
HV   Hyperstriatum ventrale
ICT   Nucleus intercalatus thalami
LAD   Lamina archistriatalis dorsalis
LH   Lamina hyperstriatica
LHy   Regio lateralis hypothalami (Lateral hypothalamic area)
LMD   Lamina medullaris dorsalis
N   Neostriatum
nTSM Nucleus tractus septomesencephalicus (Nucleus superficialis parvocellularis)
OM   Tractus occipitomesencephalicus
PA   Paleostriatum augmentatum (Caudate putamen)
PCVL Plexus choroideus ventriculi lateralis (Choroid plexus within lateral ventricle)
PHN   Nucleus periventricularis hypothalami
PVN   Nucleus paraventricularis magnocellularis (Paraventricular nucleus)
QF   Tractus quintofrontalis
ROT   Nucleus rotundus
RSd   Nucleus reticularis superior, pars dorsalis
SMe   Stria medullaris
Tn   Nucleus taeniae
TrO   Tractus opticus
TSM   Tractus septomesencephalicus
VL   Ventriculus lateralis
VMN   Nucleus ventromedialis hypothalami

AId   Archistriatum intermedium, pars dorsalis (Zeier and Karten)
AIv   Archistriatum intermedium, pars ventralis (Zeier and Karten)
AL   Ansa lenticularis
ALA   Nucleus ansae lenticularis anterior [rostralis]
Am   Archistriatum mediale (Zeier and Karten)
APH   Area parahippocampalis
CDL   Area corticoidea dorsolateralis
CHCS Tractus cortico-habenularis et cortico-septalis
CIO   Capsula interna occipitalis
CO   Chiasma opticum
CPi   Cortex piriformis
DA   Tractus dorso-archistriaticus
DLAI Nucleus dorsolateralis anterior [rostralis] thalami, pars lateralis
DLAm Nucleus dorsolateralis anterior [rostralis] thalami, pars medialis
DMA   Nucleus dorsomedialis anterior [rostralis] thalami
DSD   Decussatio supraoptica dorsalis
DSV   Decussatio supraoptica ventralis
FL   Field L
FPL   Fasciculus prosencephali lateralis (Lateral forebrain bundle)
FPM   Fasciculus prosencephali medialis (Medial forebrain bundle)
GLv   Nucleus geniculatus lateralis, pars ventralis
HM   Nucleus habenularis medialis
Hp   Hippocampus
HV   Hyperstriatum ventrale
ICT   Nucleus intercalatus thalami
LAD   Lamina archistriatalis dorsalis
LH   Lamina hyperstriatica
LHy   Regio lateralis hypothalami (Lateral hypothalamic area)
LMD   Lamina medullaris dorsalis
N   Neostriatum
nTSM Nucleus tractus septomesencephalicus (Nucleus superficialis parvocellularis)
OM   Tractus occipitomesencephalicus
PA   Paleostriatum augmentatum (Caudate putamen)
PCVL Plexus choroideus ventriculi lateralis (Choroid plexus within lateral ventricle)
PHN   Nucleus periventricularis hypothalami
PVN   Nucleus paraventricularis magnocellularis (Paraventricular nucleus)
QF   Tractus quintofrontalis
ROT   Nucleus rotundus
RSd   Nucleus reticularis superior, pars dorsalis
SMe   Stria medullaris
Tn   Nucleus taeniae
TrO   Tractus opticus
TSM   Tractus septomesencephalicus
VL   Ventriculus lateralis
VMN   Nucleus ventromedialis hypothalami

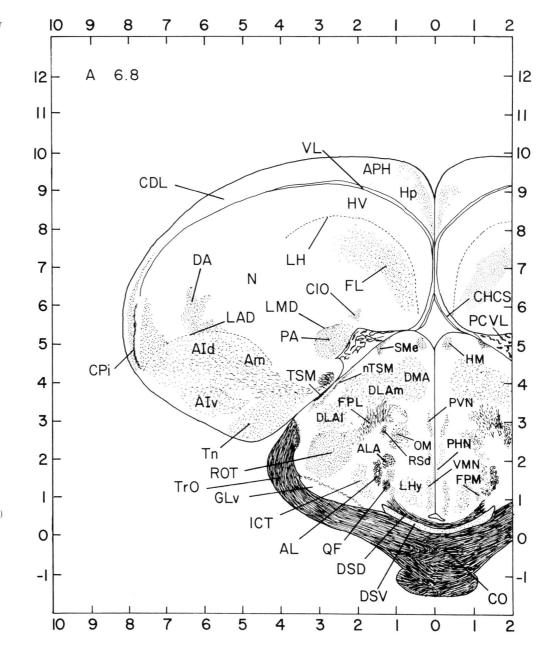

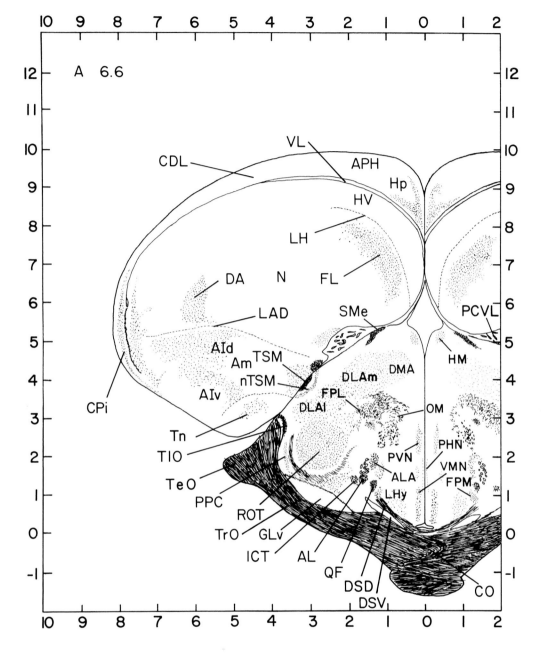

AId Archistriatum intermedium, pars dorsalis (Zeier and Karten)
AIv Archistriatum intermedium, pars ventralis (Zeier and Karten)
AL Ansa lenticularis
ALA Nucleus ansae lenticularis anterior [rostralis]
Am Archistriatum mediale (Zeier and Karten)
APH Area parahippocampalis
CDL Area corticoidea dorsolateralis
CO Chiasma opticum
CPi Cortex piriformis
DA Tractus dorso-archistriaticus
DLAl Nucleus dorsolateralis anterior [rostralis] thalami, pars lateralis
DLAm Nucleus dorsolateralis anterior [rostralis] thalami, pars medialis
DMA Nucleus dorsomedialis anterior [rostralis] thalami
DSD Decussatio supraoptica dorsalis
DSV Decussatio supraoptica ventralis
FL Field L
FPL Fasciculus prosencephali lateralis (Lateral forebrain bundle)
FPM Fasciculus prosencephali medialis (Medial forebrain bundle)
GLv Nucleus geniculatus lateralis, pars ventralis
HM Nucleus habenularis medialis
Hp Hippocampus
HV Hyperstriatum ventrale
ICT Nucleus intercalatus thalami
LAD Lamina archistriatalis dorsalis
LH Lamina hyperstriatica
LHy Regio lateralis hypothalami (Lateral hypothalamic area)
N Neostriatum
nTSM Nucleus tractus septomesencephalicus (Nucleus superficialis parvocellularis)
OM Tractus occipitomesencephalicus
PCVL Plexus choroideus ventriculi lateralis (Choroid plexus within lateral ventricle)
PHN Nucleus periventricularis hypothalami
PPC Nucleus principalis precommissuralis
PVN Nucleus paraventricularis magnocellularis (Paraventricular nucleus)
QF Tractus quintofrontalis
ROT Nucleus rotundus
SMe Stria medullaris
TeO Tectum opticum
TIO Tractus isthmo-opticus
Tn Nucleus taeniae
TrO Tractus opticus
TSM Tractus septomesencephalicus
VL Ventriculus lateralis
VMN Nucleus ventromedialis hypothalami

AId Archistriatum intermedium, pars dorsalis (Zeier
      and Karten)
AIv Archistriatum intermedium, pars ventralis
      (Zeier and Karten)
AL   Ansa lenticularis
ALA Nucleus ansae lenticularis anterior [rostralis]
Am   Archistriatum mediale (Zeier and Karten)
APH Area parahippocampalis
CDL Area corticoidea dorsolateralis
CO   Chiasma opticum
CPi Cortex piriformis
DA   Tractus dorso-archistriaticus
DLA Nucleus dorsolateralis anterior [rostralis]
      thalami
DMA Nucleus dorsomedialis anterior [rostralis]
      thalami
DSD Decussatio supraoptica dorsalis
DSV Decussatio supraoptica ventralis
FL   Field L
GLv Nucleus geniculatus lateralis, pars ventralis
HL   Nucleus habenularis lateralis
HM   Nucleus habenularis medialis
Hp   Hippocampus
HV   Hyperstriatum ventrale
ICT Nucleus intercalatus thalami
LAD Lamina archistriatalis dorsalis
LH   Lamina hyperstriatica
LHy Regio lateralis hypothalami (Lateral
      hypothalamic area)
LMmc Nucleus lentiformis mesencephali, pars
      magnocellularis
N   Neostriatum
nTSM Nucleus tractus septomesencephalicus
      (Nucleus superficialis parvocellularis)
OM   Tractus occipitomesencephalicus
OV   Nucleus ovoidalis
PCVL Plexus choroideus ventriculi lateralis (Choroid
      plexus within lateral ventricle)
PHN Nucleus periventricularis hypothalami
PPC Nucleus principalis precommissuralis
PV   Nucleus posteroventralis thalami (Kuhlenbeck)
PVN Nucleus paraventricularis magnocellularis
      (Paraventricular nucleus)
QF   Tractus quintofrontalis
ROT Nucleus rotundus
SMe Stria medullaris
SRt Nucleus subrotundus
T   Nucleus triangularis
TeO Tectum opticum
TIO Tractus isthmo-opticus
Tn   Nucleus taeniae
TrO Tractus opticus
TSM Tractus septomesencephalicus
TT   Tractus tectothalamicus
TTS Tractus thalamostriaticus
VL   Ventriculus lateralis
VMN Nucleus ventromedialis hypothalami

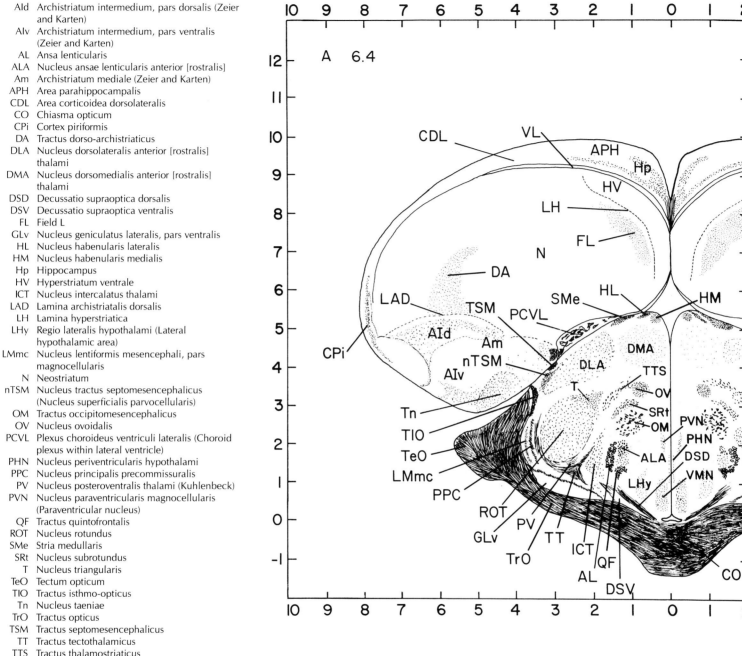

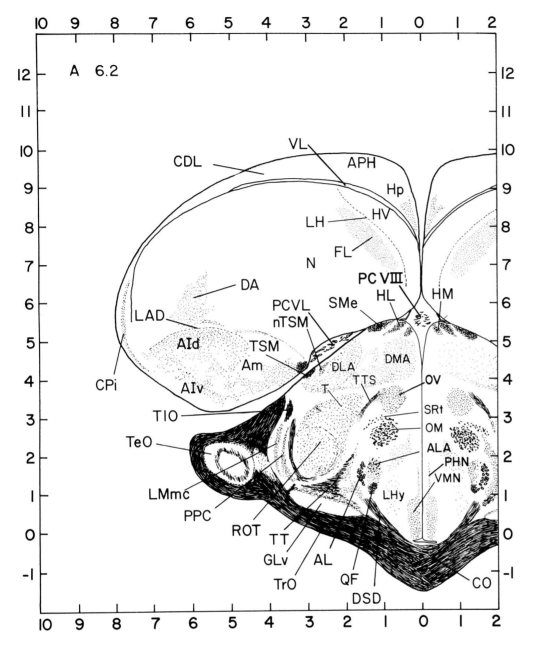

AId Archistriatum intermedium, pars dorsalis
     (Zeier and Karten)
AIv Archistriatum intermedium, pars ventralis
     (Zeier and Karten)
AL   Ansa lenticularis
ALA Nucleus ansae lenticularis anterior [rostralis]
Am   Archistriatum mediale (Zeier and Karten)
APH Area parahippocampalis
CDL Area corticoidea dorsolateralis
CO   Chiasma opticum
CPi Cortex piriformis
DA   Tractus dorso-archistriaticus
DLA Nucleus dorsolateralis anterior [rostralis]
     thalami
DMA Nucleus dorsomedialis anterior [rostralis]
     thalami
DSD Decussatio supraoptica dorsalis
FL   Field L
GLv Nucleus geniculatus lateralis, pars ventralis
HL   Nucleus habenularis lateralis
HM   Nucleus habenularis medialis
Hp   Hippocampus
HV   Hyperstriatum ventrale
LAD Lamina archistriatalis dorsalis (Zeier and
     Karten)
LH   Lamina hyperstriatica
LHy Regio lateralis hypothalami (Lateral
     hypothalamic area)
LMmc Nucleus lentiformis mesencephali, pars
     magnocellularis
N   Neostriatum
nTSM Nucleus tractus septomesencephalicus
     (Nucleus superficialis parvocellularis)
OM   Tractus occipitomesencephalicus
OV   Nucleus ovoidalis
PCV III Plexus choroideus ventriculi tertii (Choroid
     plexus within third ventricle)
PCVL Plexus choroideus ventriculi lateralis (Choroid
     plexus within lateral ventricle)
PHN Nucleus periventricularis hypothalami
PPC Nucleus principalis precommissuralis
QF   Tractus quintofrontalis
ROT Nucleus rotundus
SMe Stria medullaris
SRt Nucleus subrotundus
T   Nucleus triangularis
TeO Tectum opticum
TIO Tractus isthmo-opticus
TrO Tractus opticus
TSM Tractus septomesencephalicus
TT   Tractus tectothalamicus
TTS Tractus thalamostriaticus
VL   Ventriculus lateralis
VMN Nucleus ventromedialis hypothalami

AId   Archistriatum intermedium, pars dorsalis
        (Zeier and Karten)
AIv   Archistriatum intermedium, pars ventralis
        (Zeier and Karten)
AL   Ansa lenticularis
APH   Area parahippocampalis
CDL   Area corticoidea dorsolateralis
CO   Chiasma opticum
CPi   Cortex piriformis
DA   Tractus dorso-archistriaticus
DLP   Nucleus dorsolateralis posterior [caudalis]
        thalami
DMP   Nucleus dorsomedialis posterior [caudalis]
        thalami
FL   Field L
GLdp Nucleus geniculatus lateralis, pars dorsalis
        principalis
GLv   Nucleus geniculatus lateralis, pars ventralis
HL   Nucleus habenularis lateralis
HM   Nucleus habenularis medialis
Hp   Hippocampus
HV   Hyperstriatum ventrale
LAD   Lamina archistriatalis dorsalis
LH   Lamina hyperstriatica
LHy   Regio lateralis hypothalami (Lateral
        hypothalamic area)
LMmc Nucleus lentiformis mesencephali, pars
        magnocellularis
LMpc Nucleus lentiformis mesencephali, pars
        parvocellularis
N   Neostriatum
nTSM Nucleus tractus septomesencephalicus
        (Nucleus superficialis parvocellularis)
OM   Tractus occipitomesencephalicus
OV   Nucleus ovoidalis
PCV III Plexus choroideus ventriculi tertii (Choroid
        plexus within third ventricle)
PHN   Nucleus periventricularis hypothalami
PPC   Nucleus principalis precommissuralis
PTD   Nucleus pretectalis diffusus
PVO   Organum paraventriculare (Paraventricular
        organ)
QF   Tractus quintofrontalis
ROT   Nucleus rotundus
SCE   Stratum cellulare externum
SGFS Stratum griseum et fibrosum superficiale
SMe   Stria medullaris
SO   Stratum opticum
SRt   Nucleus subrotundus
T   Nucleus triangularis
TIO   Tractus isthmo-opticus
TrO   Tractus opticus
TSM   Tractus septomesencephalicus
TT   Tractus tectothalamicus
TTS   Tractus thalamostriaticus
VL   Ventriculus lateralis
VMN   Nucleus ventromedialis hypothalami

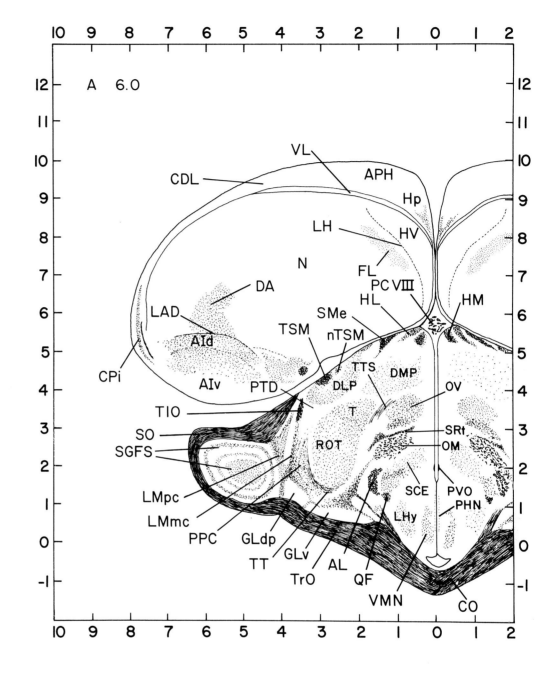

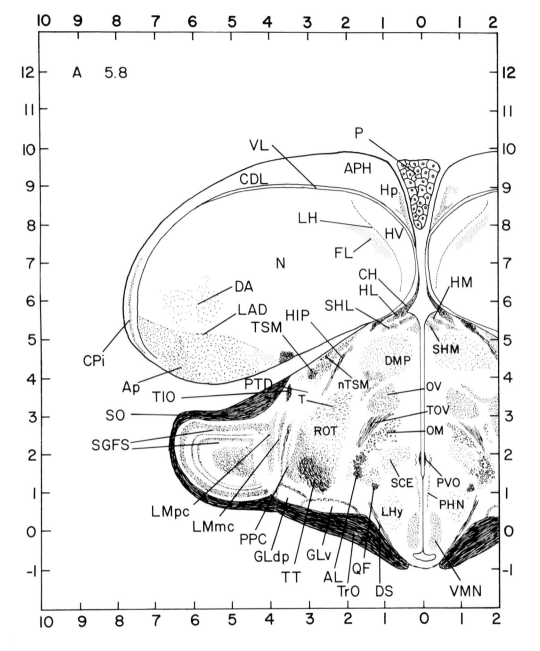

AL   Ansa lenticularis
Ap   Archistriatum posterior [caudale] (Zeier and
         Karten)
APH   Area parahippocampalis
CDL   Area corticoidea dorsolateralis
CH   Tractus corticohabenularis
CPi   Cortex piriformis
DA   Tractus dorso-archistriaticus
DMP   Nucleus dorsomedialis posterior [caudalis]
         thalami
DS   Decussatio supraoptica
FL   Field L
GLdp Nucleus geniculatus lateralis, pars dorsalis
         principalis
GLv   Nucleus geniculatus lateralis, pars ventralis
HIP   Tractus habenulointerpeduncularis
HL   Nucleus habenularis lateralis
HM   Nucleus habenularis medialis
Hp   Hippocampus
HV   Hyperstriatum ventrale
LAD   Lamina archistriatalis dorsalis
LH   Lamina hyperstriatica
LHy   Regio lateralis hypothalami (Lateral
         hypothalamic area)
LMmc Nucleus lentiformis mesencephali, pars
         magnocellularis
LMpc Nucleus lentiformis mesencephali, pars
         parvocellularis
N   Neostriatum
nTSM Nucleus tractus septomesencephalicus
         (Nucleus superficialis parvocellularis)
OM   Tractus occipitomesencephalicus
OV   Nucleus ovoidalis
P   Corpus pineale
PHN   Nucleus periventricularis hypothalami
PPC   Nucleus principalis precommissuralis
PTD   Nucleus pretectalis diffusus
PVO   Organum paraventriculare (Paraventricular
         organ)
QF   Tractus quintofrontalis
ROT   Nucleus rotundus
SCE   Stratum cellulare externum
SGFS Stratum griseum et fibrosum superficiale
SHL   Nucleus subhabenularis lateralis
SHM   Nucleus subhabenularis medialis
SO   Stratum opticum
T   Nucleus triangularis
TIO   Tractus isthmo-opticus
TOV   Tractus nuclei ovoidalis
TrO   Tractus opticus
TSM   Tractus septomesencephalicus
TT   Tractus tectothalamicus
VL   Ventriculus lateralis
VMN   Nucleus ventromedialis hypothalami

AL   Ansa lenticularis
Ap   Archistriatum posterior [caudale] (Zeier and
       Karten)
APH   Area parahippocampalis
CDL   Area corticoidea dorsolateralis
CPi   Cortex piriformis
DA   Tractus dorso-archistriaticus
DIP   Nucleus dorsointermedius posterior thalami
DLP   Nucleus dorsolateralis posterior [caudalis]
       thalami
DMP   Nucleus dorsomedialis posterior [caudalis]
       thalami
GLdp   Nucleus geniculatus lateralis, pars dorsalis
       principalis
GLv   Nucleus geniculatus lateralis, pars ventralis
HIP   Tractus habenulointerpeduncularis
HL   Nucleus habenularis lateralis
HM   Nucleus habenularis medialis
Hp   Hippocampus
HV   Hyperstriatum ventrale
IH   Nucleus inferioris hypothalami
IN   Nucleus infundibuli hypothalami
LH   Lamina hyperstriatica
LHy   Regio lateralis hypothalami (Lateral
       hypothalamic area)
LMmc   Nucleus lentiformis mesencephali, pars
       magnocellularis
LMpc   Nucleus lentiformis mesencephali, pars
       parvocellularis
N   Neostriatum
nTSM   Nucleus tractus septomesencephalicus
       (Nucleus superficialis parvocellularis)
OM   Tractus occipitomesencephalicus
OV   Nucleus ovoidalis
P   Corpus pineale
PHN   Nucleus periventricularis hypothalami
PPC   Nucleus principalis precommissuralis
PTM   Nucleus pretectalis medialis
PVO   Organum paraventriculare (Paraventricular
       organ)
QF   Tractus quintofrontalis
ROT   Nucleus rotundus
SCE   Stratum cellulare externum
SGC   Stratum griseum centrale
SGFS   Stratum griseum et fibrosum superficiale
SHL   Nucleus subhabenularis lateralis
SHM   Nucleus subhabenularis medialis
SO   Stratum opticum
T   Nucleus triangularis
TIO   Tractus isthmo-opticus
TOV   Tractus nuclei ovoidalis
TrO   Tractus opticus
TSM   Tractus septomesencephalicus
TT   Tractus tectothalamicus
VL   Ventriculus lateralis

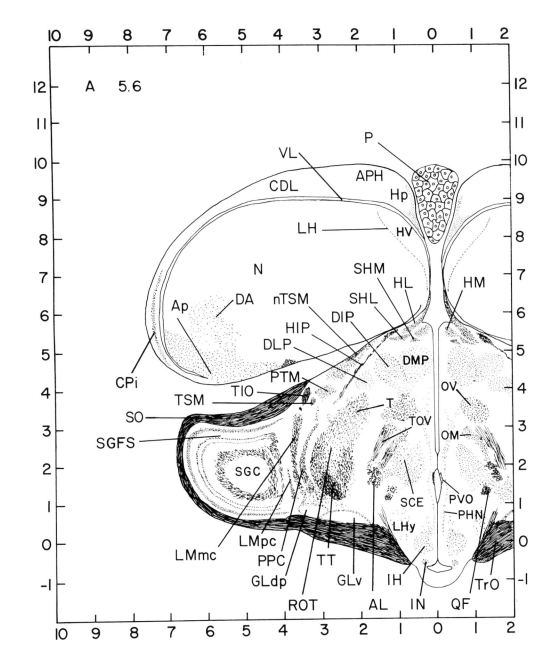

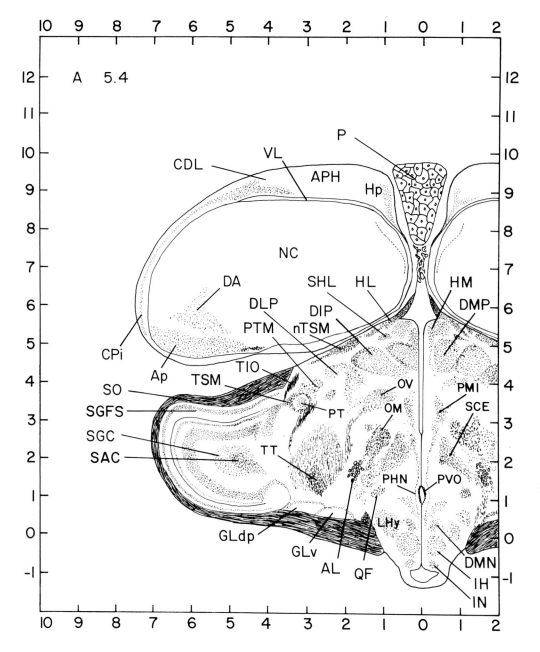

AL   Ansa lenticularis
Ap   Archistriatum posterior [caudale] (Zeier and
     Karten)
APH Area parahippocampalis
CDL Area corticoidea dorsolateralis
CPi Cortex piriformis
DA   Tractus dorso-archistriaticus
DIP Nucleus dorsointermedius posterior thalami
DLP Nucleus dorsolateralis posterior [caudalis]
     thalami
DMN Nucleus dorsomedialis hypothalami
DMP Nucleus dorsomedialis posterior [caudalis]
     thalami
GLdp Nucleus geniculatus lateralis, pars dorsalis
     principalis
GLv Nucleus geniculatus lateralis, pars ventralis
HL   Nucleus habenularis lateralis
HM   Nucleus habenularis medialis
Hp   Hippocampus
IH   Nucleus inferioris hypothalami
IN   Nucleus infundibuli hypothalami
LHy Regio lateralis hypothalami (Lateral
     hypothalamic area)
NC   Neostriatum caudale
nTSM Nucleus tractus septomesencephalicus (Nucleus
     superficialis parvocellularis)
OM   Tractus occipitomesencephalicus
OV   Nucleus ovoidalis
P   Corpus pineale
PHN Nucleus periventricularis hypothalami
PMI Nucleus paramedianus internus thalami
PT   Nucleus pretectalis
PTM Nucleus pretectalis medialis
PVO Organum paraventriculare (Paraventricular
     organ)
QF   Tractus quintofrontalis
SAC Stratum album centrale
SCE Stratum cellulare externum
SGC Stratum griseum centrale
SGFS Stratum griseum et fibrosum superficiale
SHL Nucleus subhabenularis lateralis
SO   Stratum opticum
TIO Tractus isthmo-opticus
TSM Tractus septomesencephalicus
TT   Tractus tectothalamicus
VL   Ventriculus lateralis

AL Ansa lenticularis
ALP Nucleus ansae lenticularis posterior [caudalis]
Ap Archistriatum posterior [caudale] (Zeier and Karten)
APH Area parahippocampalis
CDL Area corticoidea dorsolateralis
CPi Cortex piriformis
DIP Nucleus dorsointermedius posterior thalami
DLP Nucleus dorsolateralis posterior [caudalis] thalami
DMN Nucleus dorsomedialis hypothalami
DMP Nucleus dorsomedialis posterior [caudalis] thalami
GLdp Nucleus geniculatus lateralis, pars dorsalis principalis
GLv Nucleus geniculatus lateralis, pars ventralis
Hp Hippocampus
IH Nucleus inferioris hypothalami
IN Nucleus infundibuli hypothalami
LHy Regio lateralis hypothalami (Lateral hypothalamic area)
ME Eminentia mediana (Median eminence)
NC Neostriatum caudale
nTSM Nucleus tractus septomesencephalicus (Nucleus superficialis parvocellularis)
OM Tractus occipitomesencephalicus
P Corpus pineale
PHN Nucleus periventricularis hypothalami
PMI Nucleus paramedianus internus thalami
PT Nucleus pretectalis
PTM Nucleus pretectalis medialis
PVO Organum paraventriculare (Paraventricular organ)
QF Tractus quintofrontalis
SAC Stratum album centrale
SCE Stratum cellulare externum
SCO Organum subcommissurale (Subcommissural organ)
SGC Stratum griseum centrale
SGFS Stratum griseum et fibrosum superficiale
SGP Stratum griseum periventriculare
SO Stratum opticum
SP Nucleus subpretectalis
SpL Nucleus spiriformis lateralis
SpM Nucleus spiriformis medialis
TIO Tractus isthmo-opticus
TSM Tractus septomesencephalicus
VL Ventriculus lateralis

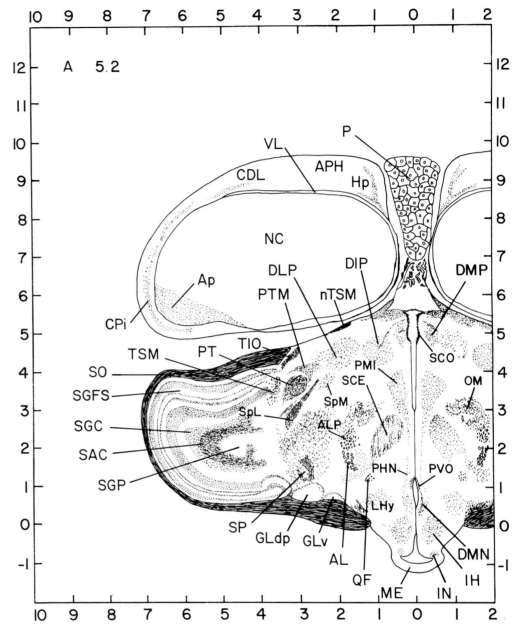

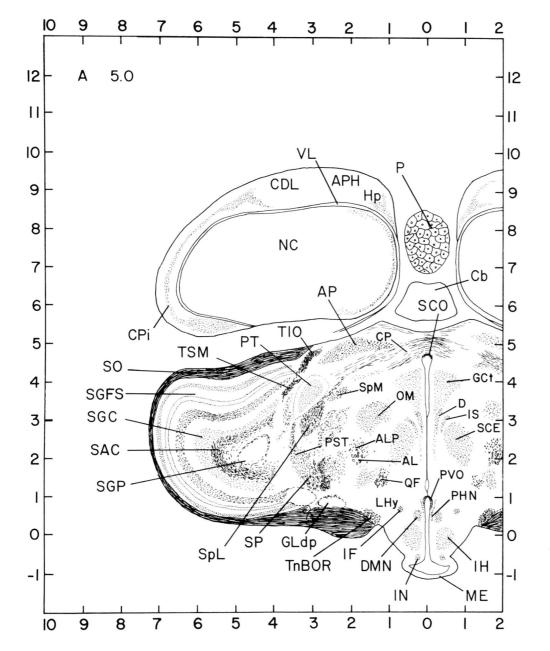

AL   Ansa lenticularis
ALP Nucleus ansae lenticularis posterior [caudalis]
AP   Area pretectalis
APH Area parahippocampalis
Cb   Cerebellum
CDL Area corticoidea dorsolateralis
CP   Commissura posterior [caudalis] (Posterior commissure)
CPi Cortex piriformis
D   Nucleus of Darkschewitsch; nucleus paragrisealis centralis mesencephali (ICAAN)
DMN Nucleus dorsomedialis hypothalami
GCt Substantia grisea centralis (Central gray)
GLdp Nucleus geniculatus lateralis, pars dorsalis principalis
Hp   Hippocampus
IF   Tractus infundibularis
IH   Nucleus inferioris hypothalami
IN   Nucleus infundibuli hypothalami
IS   Nucleus interstitialis (Cajal)
LHy Regio lateralis hypothalami (Lateral hypothalamic area)
ME   Eminentia mediana (Median eminence)
NC   Neostriatum caudale
OM   Tractus occipitomesencephalicus
P   Corpus pineale
PHN Nucleus periventricularis hypothalami
PST Tractus pretecto-subpretectalis
PT   Nucleus pretectalis
PVO Organum paraventriculare (Paraventricular organ)
QF   Tractus quintofrontalis
SAC Stratum album centrale
SCE Stratum cellulare externum
SCO Organum subcommissurale (Subcommissural organ)
SGC Stratum griseum centrale
SGFS Stratum griseum et fibrosum superficiale
SGP Stratum griseum periventriculare
SO   Stratum opticum
SP   Nucleus subpretectalis
SpL Nucleus spiriformis lateralis
SpM Nucleus spiriformis medialis
TIO Tractus isthmo-opticus
TnBOR Tractus nuclei optici basalis (Tractus nuclei ectomamillaris; tract of the basal optic root)
TSM Tractus septomesencephalicus
VL   Ventriculus lateralis

AL   Ansa lenticularis
ALP Nucleus ansae lenticularis posterior [caudalis]
APH Area parahippocampalis
Cb   Cerebellum
CDL Area corticoidea dorsolateralis
CP   Commissura posterior [caudalis] (Posterior commissure)
CPi Cortex piriformis
CT   Commissura tectalis
D   Nucleus of Darkschewitsch; nucleus paragrisealis centralis mesencephali (ICAAN)
GCt Substantia grisea centralis (Central gray)
Hp   Hippocampus
IF   Tractus infundibularis
IH   Nucleus inferioris hypothalami
IN   Nucleus infundibuli hypothalami
IPS Nucleus interstitio-pretecto-subpretectalis
IS   Nucleus interstitialis (Cajal)
LHy Regio lateralis hypothalami (Lateral hypothalamic area)
ME   Eminentia mediana (Median eminence)
MM   Nucleus mamillaris medialis
nBOR Nucleus opticus basalis; nucleus ectomamillaris (Nucleus of the basal optic root)
NC   Neostriatum caudale
nI   Nucleus intramedialis (Huber and Crosby), nucleus c (Rendahl)
OM   Tractus occipitomesencephalicus
PST Tractus pretecto-subpretectalis
PT   Nucleus pretectalis
PVO Organum paraventriculare (Paraventricular organ)
QF   Tractus quintofrontalis
SAC Stratum album centrale
SCE Stratum cellulare externum
SCO Organum subcommissurale (Subcommissural organ)
SGC Stratum griseum centrale
SGFS Stratum griseum et fibrosum superficiale
SGP Stratum griseum periventriculare
SO   Stratum opticum
SP   Nucleus subpretectalis
SpL Nucleus spiriformis lateralis
SpM Nucleus spiriformis medialis
TIO Tractus isthmo-opticus
VL   Ventriculus lateralis
VT   Ventriculus tecti mesencephali

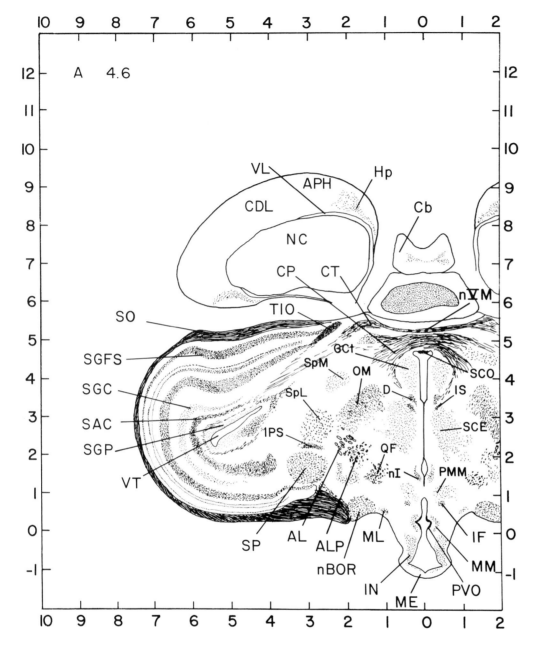

AL   Ansa lenticularis
ALP Nucleus ansae lenticularis posterior [caudalis]
APH Area parahippocampalis
Cb   Cerebellum
CDL Area corticoidea dorsolateralis
CP   Commissura posterior [caudalis] (Posterior
     commissure)
CT   Commissura tectalis
D   Nucleus of Darkschewitsch; nucleus
     paragrisealis centralis mesencephali (ICAAN)
GCt Substantia grisea centralis (Central gray)
Hp   Hippocampus
IF   Tractus infundibularis
IN   Nucleus infundibuli hypothalami
IPS Nucleus interstitio-precto-subpretectalis
IS   Nucleus interstitialis (Cajal)
ME   Eminentia mediana (Median eminence)
ML   Nucleus mamillaris lateralis
MM   Nucleus mamillaris medialis
nBOR Nucleus opticus basalis; nucleus
     ectomamillaris (Nucleus of the basal optic
     root)
n V M Nucleus mesencephalicus nervi trigemini
NC   Neostriatum caudale
nI   Nucleus intramedialis (Huber and Crosby),
     nucleus c (Rendahl)
OM   Tractus occipitomesencephalicus
PMM Nucleus premamillaris
PVO Organum paraventriculare (Paraventricular
     organ)
QF   Tractus quintofrontalis
SAC Stratum album centrale
SCE Stratum cellulare externum
SCO Organum subcommissurale (Subcommissural
     organ)
SGC Stratum griseum centrale
SGFS Stratum griseum et fibrosum superficiale
SGP Stratum griseum periventriculare
SO   Stratum opticum
SP   Nucleus subpretectalis
SpL Nucleus spiriformis lateralis
SpM Nucleus spiriformis medialis
TIO Tractus isthmo-opticus
VL   Ventriculus lateralis
VT   Ventriculus tecti mesencephali

A 4.6 ENLARGEMENT OF OPTIC TECTUM*

Systems of nomenclature for the optic tectum:

Numerical system (1–15)—Cajal (1911)

Alphabetic system (a–j)—Cowan, Adamson, and Powell (1961)

  SAC Stratum album centrale
  SGC Stratum griseum centrale
  SGFS Stratum griseum et fibrosum superficiale
  SGP Stratum griseum periventriculare
  SO Stratum opticum

* Optic tectum is also referred to as colliculus mesencephali (Cohen and Karten) and tectum mesencephali (ICAAN).

**Note that layer number 11 does not appear to be a distinct lamina in the chick. Cowan et al. (1961) likewise show layers 10 and 11 as one layer (layer i) in the pigeon. For clarification of the organization of the avian optic tectum refer to Hunt and Brecha (1984).

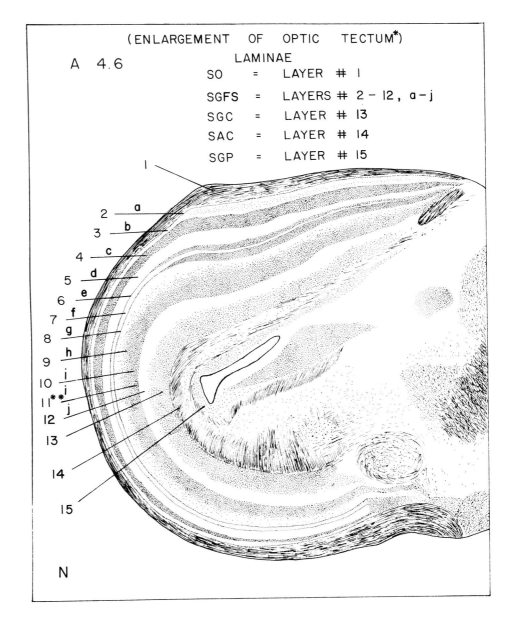

(ENLARGEMENT OF OPTIC TECTUM*)

A 4.6

LAMINAE

SO = LAYER # 1
SGFS = LAYERS # 2 – 12, a–j
SGC = LAYER # 13
SAC = LAYER # 14
SGP = LAYER # 15

N

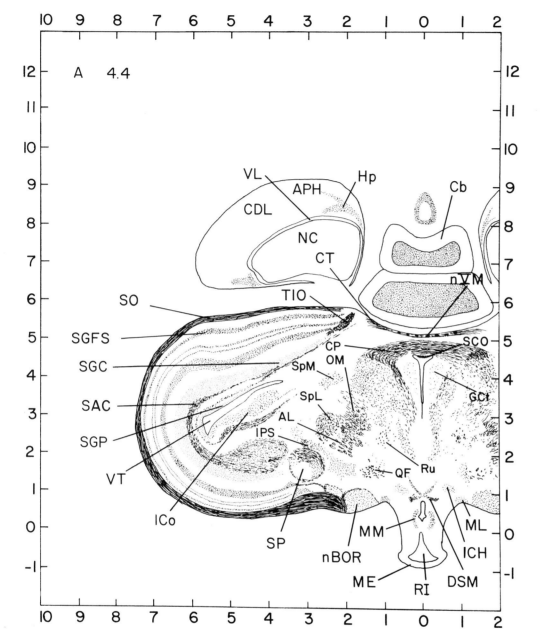

AL	Ansa lenticularis
APH	Area parahippocampalis
Cb	Cerebellum
CDL	Area corticoidea dorsolateralis
CP	Commissura posterior [caudalis] (Posterior commissure)
CT	Commissura tectalis
DSM	Decussatio supramamillaris
GCt	Substantia grisea centralis (Central gray)
Hp	Hippocampus
ICH	Nucleus intercalatus hypothalami
ICo	Nucleus intercollicularis
IPS	Nucleus interstitio-pretecto-subpretectalis
ME	Eminentia mediana (Median eminence)
ML	Nucleus mamillaris lateralis
MM	Nucleus mamillaris medialis
nBOR	Nucleus opticus basalis; nucleus ectomamillaris (Nucleus of the basal optic root)
n V M	Nucleus mesencephalicus nervi trigemini
NC	Neostriatum caudale
OM	Tractus occipitomesencephalicus
QF	Tractus quintofrontalis
RI	Recessus inframamillaris; recessus infundibuli (Infundibular recess)
Ru	Nucleus ruber
SAC	Stratum album centrale
SCO	Organum subcommissurale (Subcommissural organ)
SGC	Stratum griseum centrale
SGFS	Stratum griseum et fibrosum superficiale
SGP	Stratum griseum periventriculare
SO	Stratum opticum
SP	Nucleus subpretectalis
SpL	Nucleus spiriformis lateralis
SpM	Nucleus spiriformis medialis
TIO	Tractus isthmo-opticus
VL	Ventriculus lateralis
VT	Ventriculus tecti mesencephali

AL Ansa lenticularis
APH Area parahippocampalis
Cb Cerebellum
CDL Area corticoidea dorsolateralis
CT Commissura tectalis
FRM Formatio reticularis medialis mesencephali
GCt Substantia grisea centralis (Central gray)
Hp Hippocampus
ICo Nucleus intercollicularis
Imc Nucleus isthmi, pars magnocellularis
IPS Nucleus interstitio-pretecto-subpretectalis
nBOR Nucleus opticus basalis; nucleus
        ectomamillaris (Nucleus of the basal optic root)
n V M Nucleus mesencephalicus nervi trigemini
OM Tractus occipitomesencephalicus
QF Tractus quintofrontalis
RI Recessus inframamillaris; recessus infundibuli
      (Infundibular recess)
Ru Nucleus ruber
SAC Stratum album centrale
SCO Organum subcommissurale (Subcommissural
       organ)
SGC Stratum griseum centrale
SGFS Stratum griseum et fibrosum superficiale
SGP Stratum griseum periventriculare
SO Stratum opticum
SP Nucleus subpretectalis
TIO Tractus isthmo-opticus
TVM Tractus vestibulomesencephalicus (Papez)
VL Ventriculus lateralis
VT Ventriculus tecti mesencephali

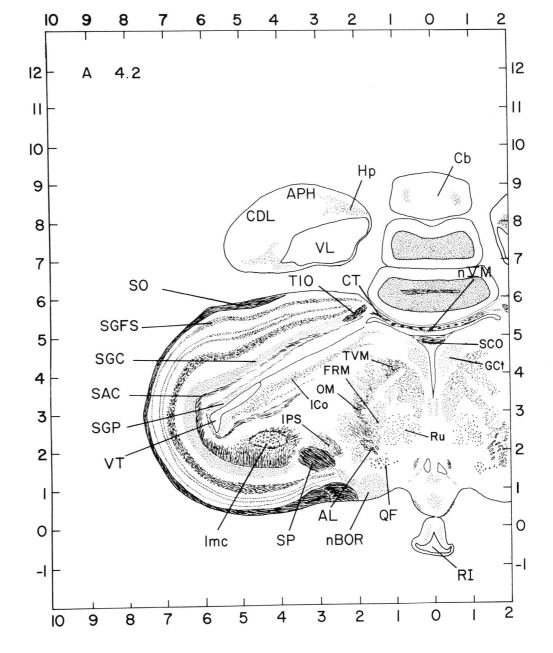

A 4.2

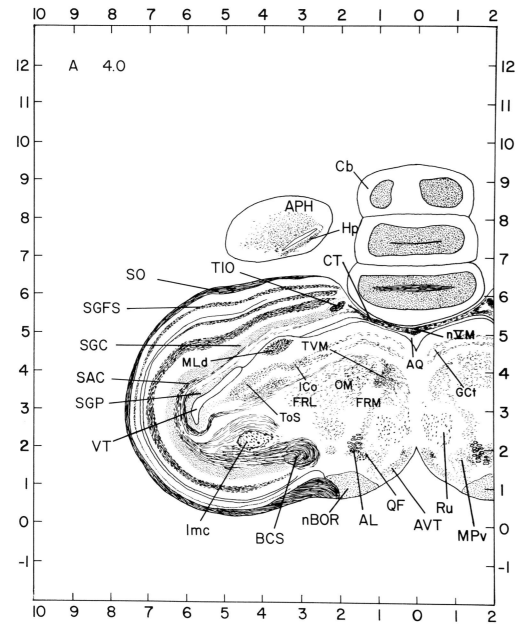

AL   Ansa lenticularis
APH Area parahippocampalis
AQ   Aqueductus mesencephali
AVT Area ventralis (Tsai)
BCS Brachium colliculi superioris
Cb   Cerebellum
CT   Commissura tectalis
FRL Formatio reticularis lateralis mesencephali
FRM Formatio reticularis medialis mesencephali
GCt Substantia grisea centralis (Central gray)
Hp   Hippocampus
ICo Nucleus intercollicularis
Imc Nucleus isthmi, pars magnocellularis
MLd Nucleus mesencephalicus lateralis, pars
       dorsalis
MPv Nucleus mesencephalicus profundus, pars
       ventralis (Jungherr)
nBOR Nucleus opticus basalis; nucleus
       ectomamillaris (Nucleus of the basal optic root)
n V M Nucleus mesencephalicus nervi trigemini
OM   Tractus occipitomesencephalicus
QF   Tractus quintofrontalis
Ru   Nucleus ruber
SAC Stratum album centrale
SGC Stratum griseum centrale
SGFS Stratum griseum et fibrosum superficiale
SGP Stratum griseum periventriculare
SO   Stratum opticum
TIO Tractus isthmo-opticus
ToS Torus semicircularis
TVM Tractus vestibulomesencephalicus (Papez)
VT   Ventriculus tecti mesencephali

AL   Ansa lenticularis
AVT Area ventralis (Tsai)
BCS Brachium colliculi superioris
Cb   Cerebellum
CT   Commissura tectalis
FLM Fasciculus longitudinalis medialis
FRL Formatio reticularis lateralis mesencephali
FRM Formatio reticularis medialis mesencephali
GCt Substantia grisea centralis (Central gray)
Imc Nucleus isthmi, pars magnocellularis
Ipc Nucleus isthmi, pars parvocellularis
MLd Nucleus mesencephalicus lateralis, pars
     dorsalis
MPv Nucleus mesencephalicus profundus, pars
     ventralis (Jungherr)
N III Nervus oculomotorius
nBOR Nucleus opticus basalis; nucleus
     ectomamillaris (Nucleus of the basal optic root)
n V M Nucleus mesencephalicus nervi trigemini
OM   Tractus occipitomesencephalicus
QF   Tractus quintofrontalis
Ru   Nucleus ruber
SAC Stratum album centrale
SGC Stratum griseum centrale
SGFS Stratum griseum et fibrosum superficiale
SGP Stratum griseum periventriculare
SO   Stratum opticum
TIO Tractus isthmo-opticus
TVM Tractus vestibulomesencephalicus (Papez)
VT   Ventriculus tecti mesencephali

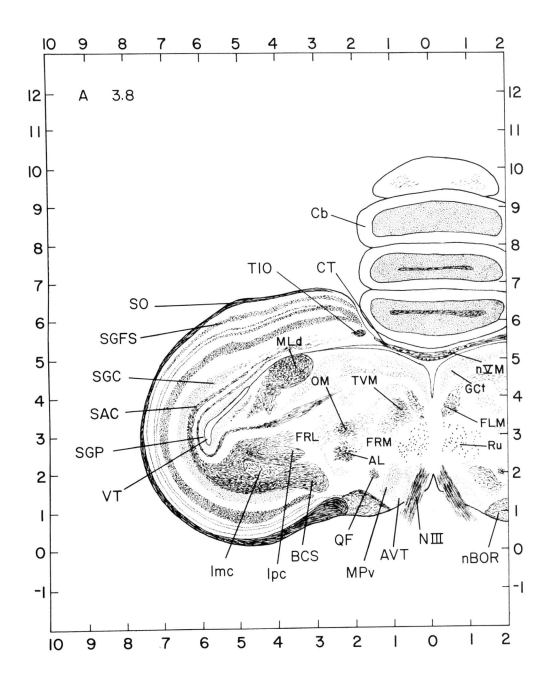

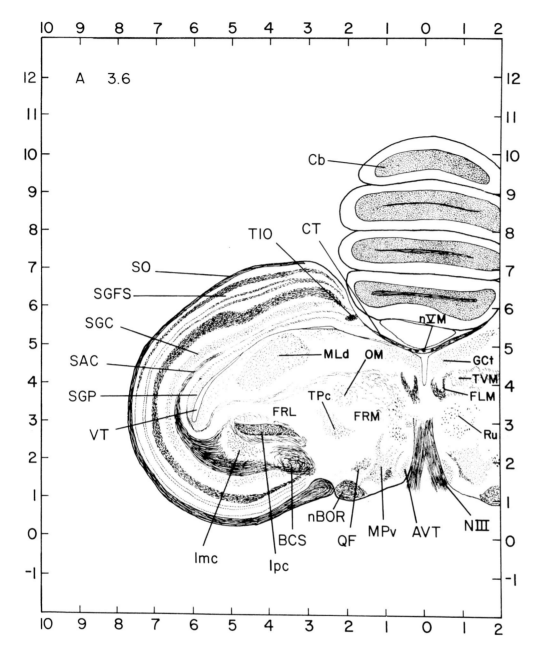

AVT   Area ventralis (Tsai)
BCS   Brachium colliculi superioris
Cb   Cerebellum
CT   Commissura tectalis
FLM   Fasciculus longitudinalis medialis
FRL   Formatio reticularis lateralis mesencephali
FRM   Formatio reticularis medialis mesencephali
GCt   Substantia grisea centralis (Central gray)
Imc   Nucleus isthmi, pars magnocellularis
Ipc   Nucleus isthmi, pars parvocellularis
MLd   Nucleus mesencephalicus lateralis, pars
      dorsalis
MPv   Nucleus mesencephalicus profundus, pars
      ventralis (Jungherr)
N III Nervus oculomotorius
nBOR Nucleus opticus basalis; nucleus
      ectomamillaris (Nucleus of the basal optic root)
n V M Nucleus mesencephalicus nervi trigemini
OM   Tractus occipitomesencephalicus
QF   Tractus quintofrontalis
Ru   Nucleus ruber
SAC   Stratum album centrale
SGC   Stratum griseum centrale
SGFS Stratum griseum et fibrosum superficiale
SGP   Stratum griseum periventriculare
SO   Stratum opticum
TIO   Tractus isthmo-opticus
TPc   Nucleus tegmenti pedunculo-pontinus, pars
      compacta (Substantia nigra)
TVM   Tractus vestibulomesencephalicus (Papez)
VT   Ventriculus tecti mesencephali

AVT Area ventralis (Tsai)
BCS Brachium colliculi superioris
Cb Cerebellum
CT Commissura tectalis
EW Nucleus of Edinger-Westphal; nucleus nervi
 oculomotorii, pars accessoria (ICAAN)
FLM Fasciculus longitudinalis medialis
FRL Formatio reticularis lateralis mesencephali
FRM Formatio reticularis medialis mesencephali
GCt Substantia grisea centralis (Central gray)
ICo Nucleus intercollicularis
Imc Nucleus isthmi, pars magnocellularis
IP Nucleus interpeduncularis
Ipc Nucleus isthmi, pars parvocellularis
MLd Nucleus mesencephalicus lateralis, pars
 dorsalis
MPv Nucleus mesencephalicus profundus, pars
 ventralis (Jungherr)
N III Nervus oculomotorius
nBOR Nucleus opticus basalis; nucleus
 ectomamillaris (Nucleus of the basal optic
 root)
n V M Nucleus mesencephalicus nervi trigemini
OM Tractus occipitomesencephalicus
OMdl Nucleus nervi oculomotorii pars dorsolateralis
OMdm Nucleus nervi oculomotorii, pars
 dorsomedialis
OMv Nucleus nervi oculomotorii, pars ventralis
QF Tractus quintofrontalis
Ru Nucleus ruber
SAC Stratum album centrale
SGC Stratum griseum centrale
SGFS Stratum griseum et fibrosum superficiale
SGP Stratum griseum periventriculare
SO Stratum opticum
TIO Tractus isthmo-opticus
TPc Nucleus tegmenti pedunculo-pontinus, pars
 compacta (Substantia nigra)
TVM Tractus vestibulomesencephalicus (Papez)
VT Ventriculus tecti mesencephali

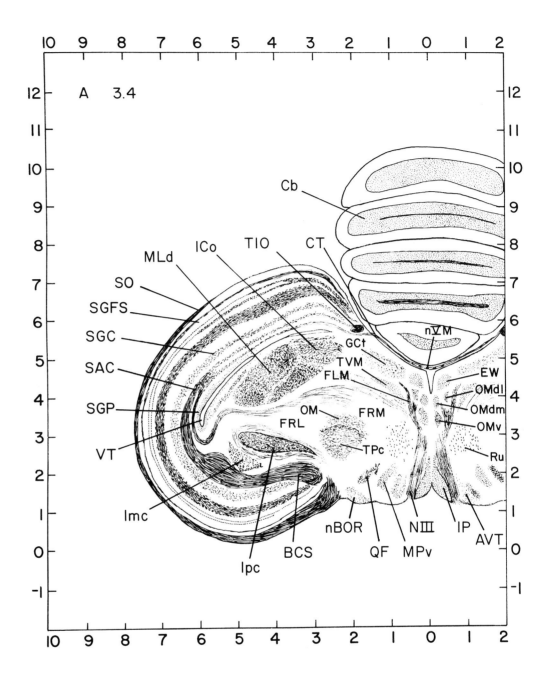

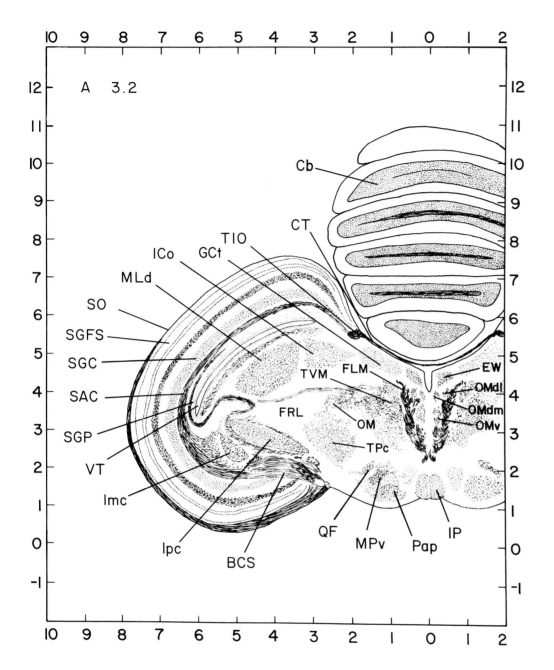

BCS   Brachium colliculi superioris
Cb   Cerebellum
CT   Commissura tectalis
EW   Nucleus of Edinger-Westphal; nucleus nervi
       oculomotorii, pars accessoria (ICAAN)
FLM   Fasciculus longitudinalis medialis
FRL   Formatio reticularis lateralis mesencephali
GCt   Substantia grisea centralis (Central gray)
ICo   Nucleus intercollicularis
Imc   Nucleus isthmi, pars magnocellularis
IP   Nucleus interpeduncularis
Ipc   Nucleus isthmi, pars parvocellularis
MLd   Nucleus mesencephalicus lateralis, pars
       dorsalis
MPv   Nucleus mesencephalicus profundus, pars
       ventralis (Jungherr)
OM   Tractus occipitomesencephalicus
OMdl   Nucleus nervi oculomotorii pars dorsolateralis
OMdm   Nucleus nervi oculomotorii, pars
       dorsomedialis
OMv   Nucleus nervi oculomotorii, pars ventralis
Pap   Nucleus papillioformis
QF   Tractus quintofrontalis
SAC   Stratum album centrale
SGC   Stratum griseum centrale
SGFS   Stratum griseum et fibrosum superficiale
SGP   Stratum griseum periventriculare
SO   Stratum opticum
TIO   Tractus isthmo-opticus
TPc   Nucleus tegmenti pedunculo-pontinus, pars
       compacta (Substantia nigra)
TVM   Tractus vestibulomesencephalicus (Papez)
VT   Ventriculus tecti mesencephali

BCA   Brachium conjunctivum ascendens
BCD   Brachium conjunctivum descendens
BCS   Brachium colliculi superioris
Cb   Cerebellum
CS   Nucleus centralis superior (Bechterew)
CT   Commissura tectalis
EW   Nucleus of Edinger-Westphal; nucleus nervi
        oculomotorii, pars accessoria (ICAAN)
FLM   Fasciculus longitudinalis medialis
FRL   Formatio reticularis lateralis mesencephali
GCt   Substantia grisea centralis (Central gray)
ICo   Nucleus intercollicularis
Imc   Nucleus isthmi, pars magnocellularis
IP   Nucleus interpeduncularis
Ipc   Nucleus isthmi, pars parvocellularis
LoC   Locus ceruleus
MLd   Nucleus mesencephalicus lateralis, pars
        dorsalis
OM   Tractus occipitomesencephalicus
OMdl Nucleus nervi oculomotorii, pars dorsolateralis
OMdm Nucleus nervi oculomotorii, pars
        dorsomedialis
OMv   Nucleus nervi oculomotorii, pars ventralis
Pap   Nucleus papilliformis
QF   Tractus quintofrontalis
SAC   Stratum album centrale
SGC   Stratum griseum centrale
SGFS Stratum griseum et fibrosum superficiale
SGP   Stratum griseum periventriculare
SO   Stratum opticum
TIO   Tractus isthmo-opticus
TPc   Nucleus tegmenti pedunculo-pontinus, pars
        compacta (Substantia nigra)
TVM   Tractus vestibulomesencephalicus (Papez)
VT   Ventriculus tecti mesencephali

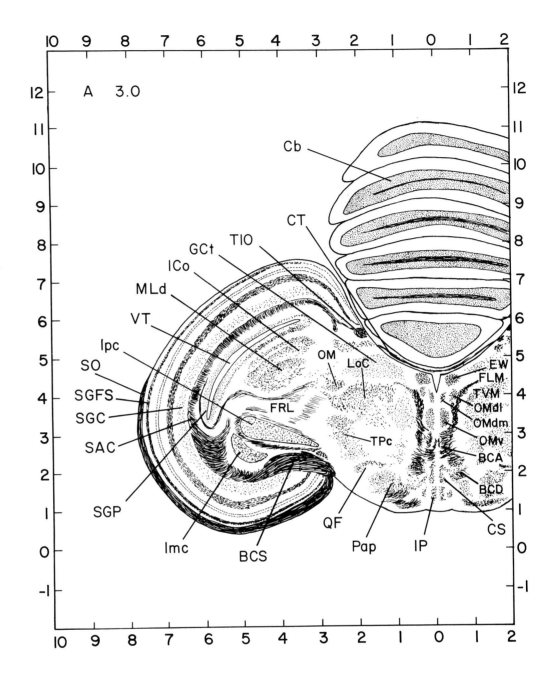

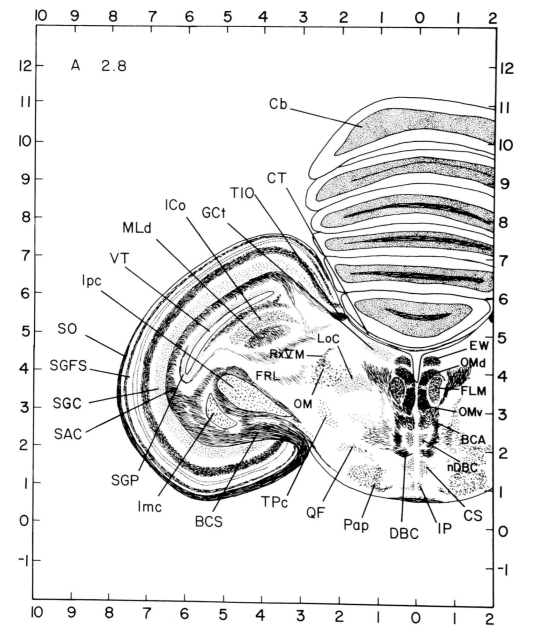

A 2.8

BCA   Brachium conjunctivum ascendens
BCS   Brachium colliculi superioris
Cb   Cerebellum
CS   Nucleus centralis superior (Bechterew)
CT   Commissura tectalis
DBC   Decussatio brachiorum conjunctivorum
EW   Nucleus of Edinger-Westphal; nucleus nervi
      oculomotorii, pars accessoria (ICAAN)
FLM   Fasciculus longitudinalis medialis
FRL   Formatio reticularis lateralis mesencephali
GCt   Substantia grisea centralis (Central gray)
ICo   Nucleus intercollicularis
Imc   Nucleus isthmi, pars magnocellularis
IP   Nucleus interpeduncularis
Ipc   Nucleus isthmi, pars parvocellularis
LoC   Locus ceruleus
MLd   Nucleus mesencephalicus lateralis, pars
      dorsalis
nDBC Nucleus decussationis brachiorum
      conjunctivorum
OM   Tractus occipitomesencephalicus
OMd   Nucleus nervi oculomotorii, pars dorsalis
OMv   Nucleus nervi oculomotorii, pars ventralis
Pap   Nucleus papillioformis
QF   Tractus quintofrontalis
Rx V M Radix mesencephalica nervi trigemini
SAC   Stratum album centrale
SGC   Stratum griseum centrale
SGFS Stratum griseum et fibrosum superficiale
SGP   Stratum griseum periventriculare
SO   Stratum opticum
TIO   Tractus isthmo-opticus
TPc   Nucleus tegmenti pedunculo-pontinus, pars
      compacta (Substantia nigra)
VT   Ventriculus tecti mesencephali

BC   Brachium conjunctivum
BCA   Brachium conjunctivum ascendens
BCS   Brachium colliculi superioris
Cb   Cerebellum
CS   Nucleus centralis superior (Bechterew)
CT   Commissura tectalis
FLM   Fasciculus longitudinalis medialis
GCt   Substantia grisea centralis (Central gray)
ICo   Nucleus intercollicularis
Imc   Nucleus isthmi, pars magnocellularis
Ipc   Nucleus isthmi, pars parvocellularis
LoC   Locus ceruleus
nDBC   Nucleus decussationis brachiorum
            conjunctivorum
OM   Tractus occipitomesencephalicus
OMd   Nucleus nervi oculomotorii, pars dorsalis
OMv   Nucleus nervi oculomotorii, pars ventralis
Pap   Nucleus papillioformis
RPO   Nucleus reticularis pontis oralis
Rx V M Radix mesencephalica nervi trigemini
SAC   Stratum album centrale
SCv   Nucleus subceruleus ventralis
SGC   Stratum griseum centrale
SGFS   Stratum griseum et fibrosum superficiale
SGP   Stratum griseum periventriculare
SLu   Nucleus semilunaris
SO   Stratum opticum
TIC   Tractus isthmocerebellaris
TIO   Tractus isthmo-opticus
VT   Ventriculus tecti mesencephali

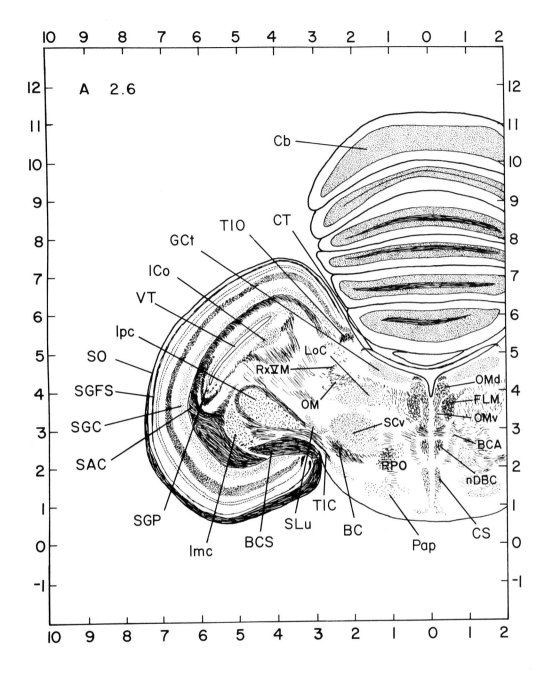

A 2.6

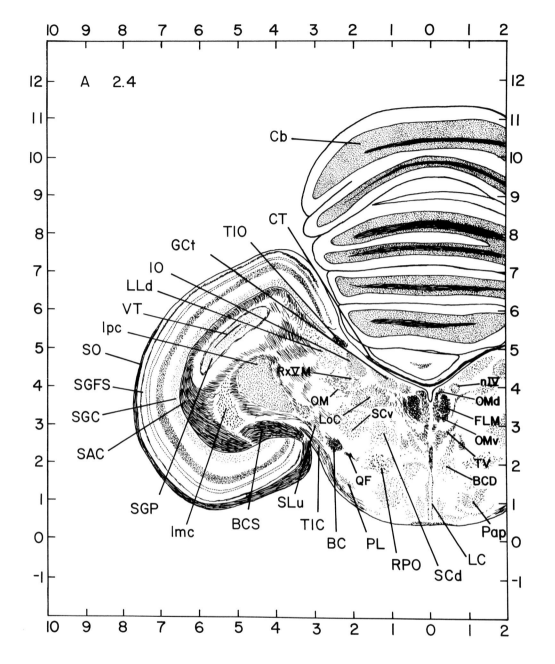

BC   Brachium conjunctivum
BCD   Brachium conjunctivum descendens
BCS   Brachium colliculi superioris
Cb   Cerebellum
CT   Commissura tectalis
FLM   Fasciculus longitudinalis medialis
GCt   Substantia grisea centralis (Central gray)
Imc   Nucleus isthmi, pars magnocellularis
IO   Nucleus isthmo-opticus
Ipc   Nucleus isthmi, pars parvocellularis
LC   Nucleus linearis caudalis
LLd   Nucleus lemnisci lateralis, pars dorsalis
          (Groebbels)
LoC   Locus ceruleus
n IV Nucleus nervi trochlearis
OM   Tractus occipitomesencephalicus
OMd   Nucleus nervi oculomotorii, pars dorsalis
OMv   Nucleus nervi oculomotorii, pars ventralis
Pap   Nucleus papillioformis
PL   Nucleus pontis lateralis
QF   Tractus quintofrontalis
RPO   Nucleus reticularis pontis oralis
Rx V M Radix mesencephalica nervi trigemini
SAC   Stratum album centrale
SCd   Nucleus subceruleus dorsalis
SCv   Nucleus subceruleus ventralis
SGC   Stratum griseum centrale
SGFS Stratum griseum et fibrosum superficiale
SGP   Stratum griseum periventriculare
SLu   Nucleus semilunaris
SO   Stratum opticum
TIC   Tractus isthmocerebellaris
TIO   Tractus isthmo-opticus
TV   Nucleus tegmenti ventralis (Gudden)
VT   Ventriculus tecti mesencephali

BC   Brachium conjunctivum
BCD   Brachium conjunctivum descendens
BCS   Brachium colliculi superioris
Cb   Cerebellum
D IV Decussatio nervi trochlearis
FLM   Fasciculus longitudinalis medialis
GCt   Substantia grisea centralis (Central gray)
Imc   Nucleus isthmi, pars magnocellularis
IO   Nucleus isthmo-opticus
Ipc   Nucleus isthmi, pars parvocellularis
LC   Nucleus linearis caudalis
LLd   Nucleus lemnisci lateralis, pars dorsalis
         (Groebbels)
LoC   Locus ceruleus
n IV Nucleus nervi trochlearis
Pap   Nucleus papillioformis
PL   Nucleus pontis lateralis
QF   Tractus quintofrontalis
RPO   Nucleus reticularis pontis oralis
SAC   Stratum album centrale
SCd   Nucleus subceruleus dorsalis
SCv   Nucleus subceruleus ventralis
SGC   Stratum griseum centrale
SGFS Stratum griseum et fibrosum superficiale
SGP   Stratum griseum periventriculare
SLu   Nucleus semilunaris
SO   Stratum opticum
STO   Organum subtrochleare (Subtrochlear organ)
TIC   Tractus isthmocerebellaris
TIO   Tractus isthmo-opticus
TV   Nucleus tegmenti ventralis (Gudden)

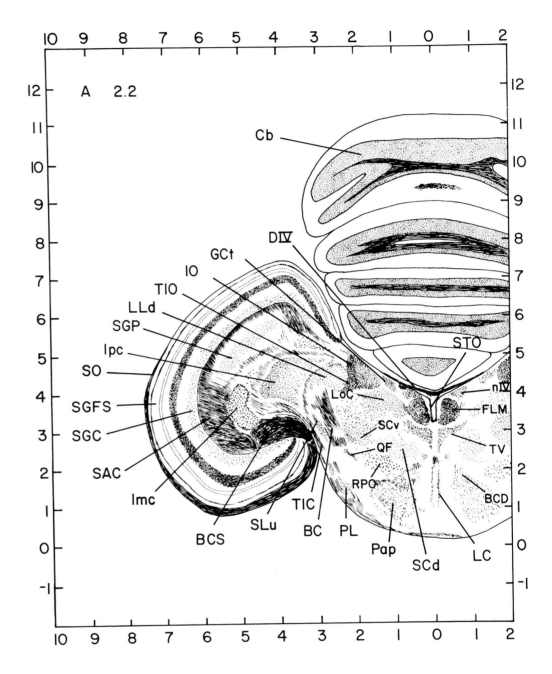

A 2.2

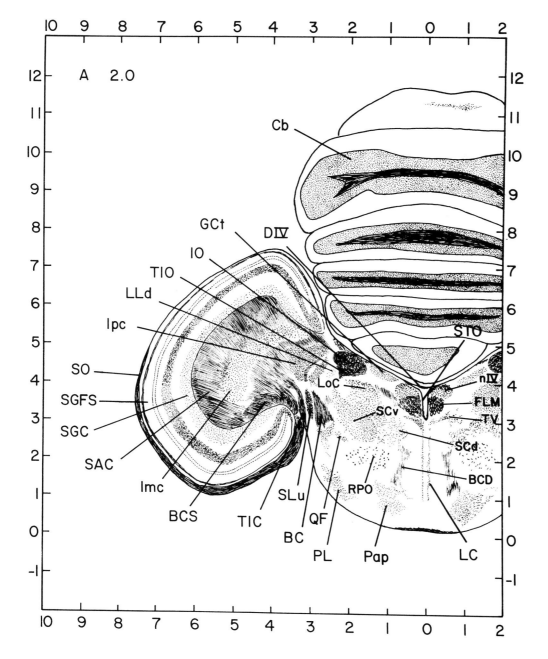

A 2.0

BC   Brachium conjunctivum
BCD Brachium conjunctivum descendens
BCS Brachium colliculi superioris
Cb   Cerebellum
D IV Decussatio nervi trochlearis
FLM Fasciculus longitudinalis medialis
GCt Substantia grisea centralis (Central gray)
Imc Nucleus isthmi, pars magnocellularis
IO   Nucleus isthmo-opticus
Ipc Nucleus isthmi, pars parvocellularis
LC   Nucleus linearis caudalis
LLd Nucleus lemnisci lateralis, pars dorsalis
     (Groebbels)
LoC Locus ceruleus
n IV Nucleus nervi trochlearis
Pap Nucleus papilliformis
PL   Nucleus pontis lateralis
QF   Tractus quintofrontalis
RPO Nucleus reticularis pontis oralis
SAC Stratum album centrale
SCd Nucleus subceruleus dorsalis
SCv Nucleus subceruleus ventralis
SGC Stratum griseum centrale
SGFS Stratum griseum et fibrosum superficiale
SLu Nucleus semilunaris
SO   Stratum opticum
STO Organum subtrochleare (Subtrochlear organ)
TIC Tractus isthmocerebellaris
TIO Tractus isthmo-opticus
TV   Nucleus tegmenti ventralis (Gudden)

BC   Brachium conjunctivum
BCD Brachium conjunctivum descendens
Cb   Cerebellum
CTz Corpus trapezoideum (Papez)
D IV Decussatio nervi trochlearis
FLM Fasciculus longitudinalis medialis
GCt Substantia grisea centralis (Central gray)
Imc Nucleus isthmi, pars magnocellularis
IO   Nucleus isthmo-opticus
LC   Nucleus linearis caudalis
LLd Nucleus lemnisci lateralis, pars dorsalis
        (Groebbels)
LLi   Nucleus lemnisci lateralis, pars intermedia
        (Arends and Zeigler); nucleus lemnisci later-
        alis, pars lateroventralis (Boord); nucleus ven-
        tralis lemnisci lateralis (Karten and Hodos)
LLv Nucleus lemnisci lateralis, pars ventralis
        (Groebbels)
LoC Locus ceruleus
n IV Nucleus nervi trochlearis
Pap Nucleus papillioformis
PL   Nucleus pontis lateralis
RPgc Nucleus reticularis pontis caudalis, pars
        gigantocellularis
RPO Nucleus reticularis pontis oralis
Rx V M Radix mesencephalica nervi trigemini
SAC Stratum album centrale
SCd Nucleus subceruleus dorsalis
SCv Nucleus subceruleus ventralis
SGC Stratum griseum centrale
SGFS Stratum griseum et fibrosum superficiale
SLu Nucleus semilunaris
SO   Stratum opticum
STO Organum subtrochleare (Subtrochlear organ)
TV   Nucleus tegmenti ventralis (Gudden)

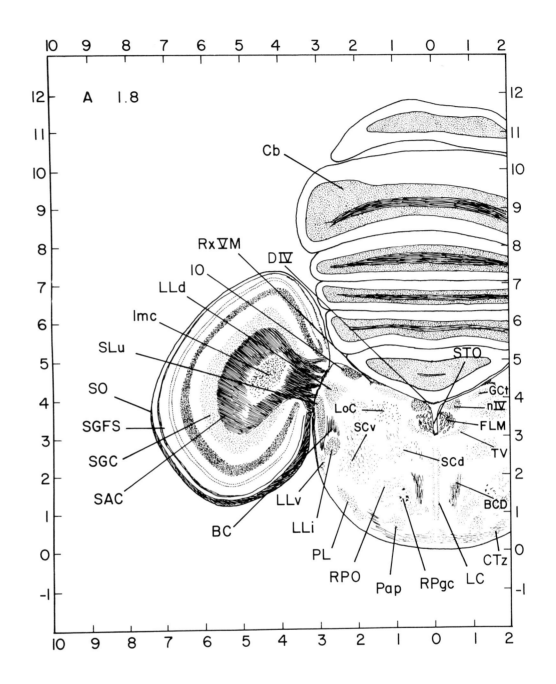

A 1.8

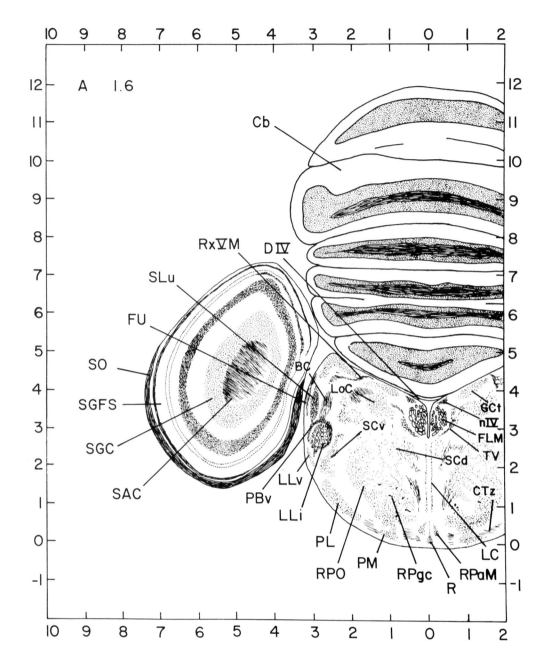

BC   Brachium conjunctivum
Cb   Cerebellum
CTz   Corpus trapezoideum (Papez)
D IV Decussatio nervi trochlearis
FLM   Fasciculus longitudinalis medialis
FU   Fasciculus uncinatus (Russell)
GCt   Substantia grisea centralis (Central gray)
LC   Nucleus linearis caudalis
LLi   Nucleus lemnisci lateralis, pars intermedia
      (Arends and Zeigler); nucleus lemnisci later-
      alis, pars lateroventralis (Boord); nucleus ven-
      tralis lemnisci lateralis (Karten and Hodos)
LLv   Nucleus lemnisci lateralis, pars ventralis
      (Groebbels)
LoC   Locus ceruleus
n IV Nucleus nervi trochlearis
PBv   Nucleus parabrachialis, pars ventralis
PL   Nucleus pontis lateralis
PM   Nucleus pontis medialis
R   Nucleus raphes (Raphe nucleus)
RPaM Nucleus reticularis paramedianus (ICAAN);
      nucleus paramedianus (Karten and Hodos)
RPgc Nucleus reticularis pontis caudalis, pars
      gigantocellularis
RPO   Nucleus reticularis pontis oralis
Rx V M Radix mesencephalica nervi trigemini
SAC   Stratum album centrale
SCd   Nucleus subceruleus dorsalis
SCv   Nucleus subceruleus ventralis
SGC   Stratum griseum centrale
SGFS Stratum griseum et fibrosum superficiale
SLu   Nucleus semilunaris
SO   Stratum opticum
TV   Nucleus tegmenti ventralis (Gudden)

BC   Brachium conjunctivum
Cb   Cerebellum
CTz Corpus trapezoideum (Papez)
D IV Decussatio nervi trochlearis
FLM Fasciculus longitudinalis medialis
FU   Fasciculus uncinatus (Russell)
LC   Nucleus linearis caudalis
LLi Nucleus lemnisci lateralis, pars intermedia
     (Arends and Zeigler); nucleus lemnisci later-
     alis, pars lateroventralis (Boord); nucleus ven-
     tralis lemnisci lateralis (Karten and Hodos)
LoC Locus ceruleus
N IV Nervus trochlearis
PBv Nucleus parabrachialis, pars ventralis
PL   Nucleus pontis lateralis
PM   Nucleus pontis medialis
R   Nucleus raphes (Raphe nucleus)
RPaM Nucleus reticularis paramedianus (ICAAN);
     nucleus paramedianus (Karten and Hodos)
RPgc Nucleus reticularis pontis caudalis, pars
     gigantocellularis
Rx V M Radix mesencephalica nervi trigemini
SCd Nucleus subceruleus dorsalis
SCv Nucleus subceruleus ventralis
SGC Stratum griseum centrale
SGFS Stratum griseum et fibrosum superficiale
SO   Stratum opticum
TD   Nucleus tegmenti dorsalis (Gudden)

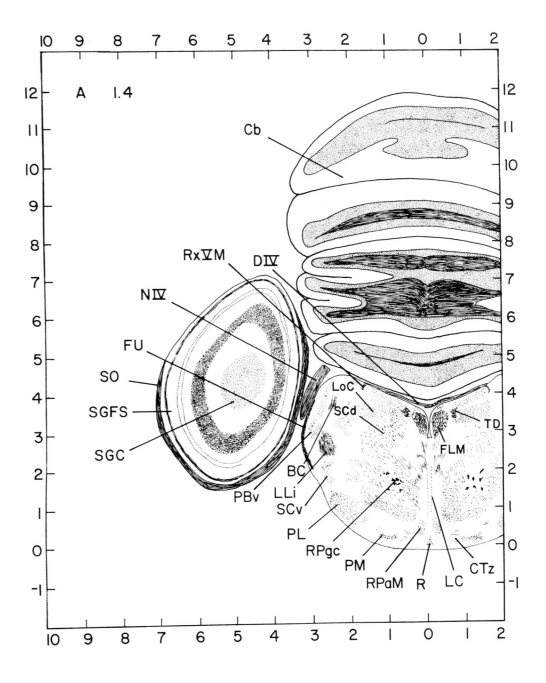

A 1.4

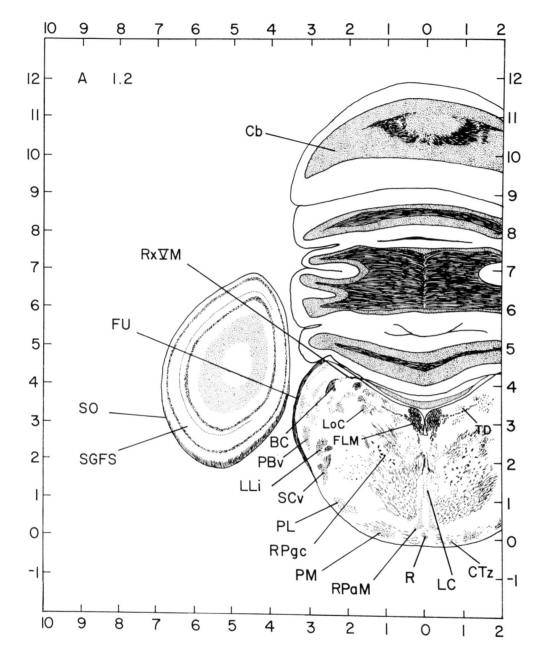

BC   Brachium conjunctivum
Cb   Cerebellum
CTz Corpus trapezoideum (Papez)
FLM Fasciculus longitudinalis medialis
FU   Fasciculus uncinatus (Russell)
LC   Nucleus linearis caudalis
LLi Nucleus lemnisci lateralis, pars intermedia
     (Arends and Zeigler); nucleus lemnisci later-
     alis, pars lateroventralis (Boord); nucleus ven-
     tralis lemnisci lateralis (Karten and Hodos)
LoC Locus ceruleus
PBv Nucleus parabrachialis, pars ventralis
PL   Nucleus pontis lateralis
PM   Nucleus pontis medialis
R   Nucleus raphes (Raphe nucleus)
RPaM Nucleus reticularis paramedianus (ICAAN);
     nucleus paramedianus (Karten and Hodos)
RPgc Nucleus reticularis pontis caudalis, pars
     gigantocellularis
Rx V M Radix mesencephalica nervi trigemini
SCv Nucleus subceruleus ventralis
SGFS Stratum griseum et fibrosum superficiale
SO   Stratum opticum
TD   Nucleus tegmenti dorsalis (Gudden)

BC Brachium conjunctivum
Cb Cerebellum
CTz Corpus trapezoideum (Papez)
FLM Fasciculus longitudinalis medialis
FU Fasciculus uncinatus (Russell)
LLi Nucleus lemnisci lateralis, pars intermedia
       (Arends and Zeigler); nucleus lemnisci lateralis,
       pars lateroventralis (Boord); nucleus ventralis
       lemnisci lateralis (Karten and Hodos)
LoC Locus ceruleus
nPr V Nucleus sensorius principalis nervi trigemini
PL Nucleus pontis lateralis
PM Nucleus pontis medialis
R Nucleus raphes (Raphe nucleus)
RPaM Nucleus reticularis paramedianus (ICAAN);
       nucleus paramedianus (Karten and Hodos)
RPgc Nucleus reticularis pontis caudalis, pars
       gigantocellularis
SCv Nucleus subceruleus ventralis
SGFS Stratum griseum et fibrosum superficiale
SO Stratum opticum
VeM Nucleus vestibularis medialis

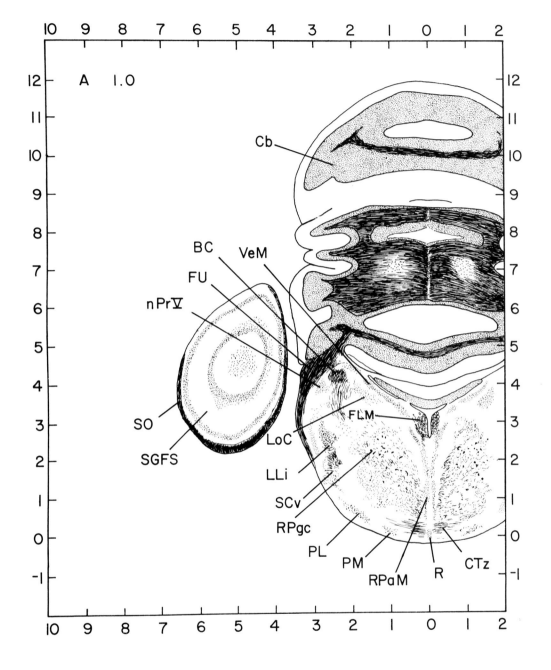

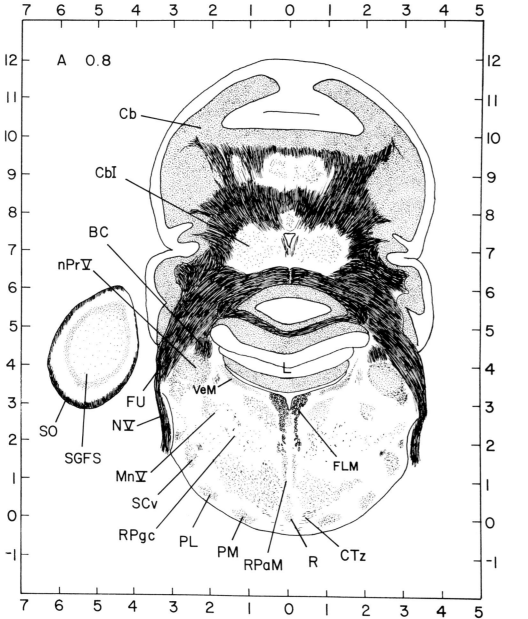

BC   Brachium conjunctivum
Cb   Cerebellum
CbI   Nucleus cerebellaris internus
CTz   Corpus trapezoideum (Papez)
FLM   Fasciculus longitudinalis medialis
FU   Fasciculus uncinatus (Russell)
L   Lingula; vinculum lingulae (ICAAN)
Mn V   Nucleus motorius nervi trigemini
N V   Nervus trigeminus
nPr V   Nucleus sensorius principalis nervi trigemini
PL   Nucleus pontis lateralis
PM   Nucleus pontis medialis
R   Nucleus raphes (Raphe nucleus)
RPaM   Nucleus reticularis paramedianus (ICAAN);
    nucleus paramedianus (Karten and Hodos)
RPgc   Nucleus reticularis pontis caudalis, pars
    gigantocellularis
SCv   Nucleus subceruleus ventralis
SGFS   Stratum griseum et fibrosum superficiale
SO   Stratum opticum
VeM   Nucleus vestibularis medialis

BC   Brachium conjunctivum
Cb   Cerebellum
Cbl   Nucleus cerebellaris internus
CTz   Corpus trapezoideum (Papez)
FLM   Fasciculus longitudinalis medialis
FU   Fasciculus uncinatus (Russell)
L   Lingula; vinculum lingulae (ICAAN)
Mn V   Nucleus motorius nervi trigemini
N V   Nervus trigeminus
nPr V Nucleus sensorius principalis nervi trigemini
PL   Nucleus pontis lateralis
PM   Nucleus pontis medialis
R   Nucleus raphes (Raphe nucleus)
RPaM   Nucleus reticularis paramedianus (ICAAN);
       nucleus paramedianus (Karten and Hodos)
RPgc   Nucleus reticularis pontis caudalis, pars
       gigantocellularis
SCv   Nucleus subceruleus ventralis
VeM   Nucleus vestibularis medialis

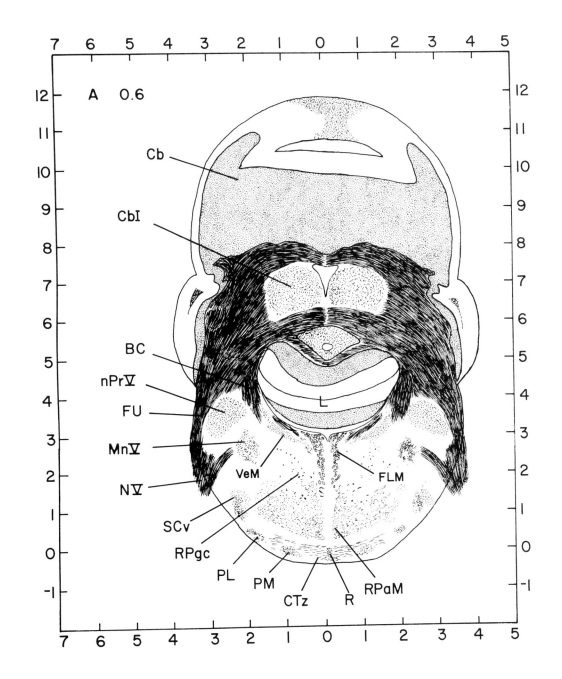

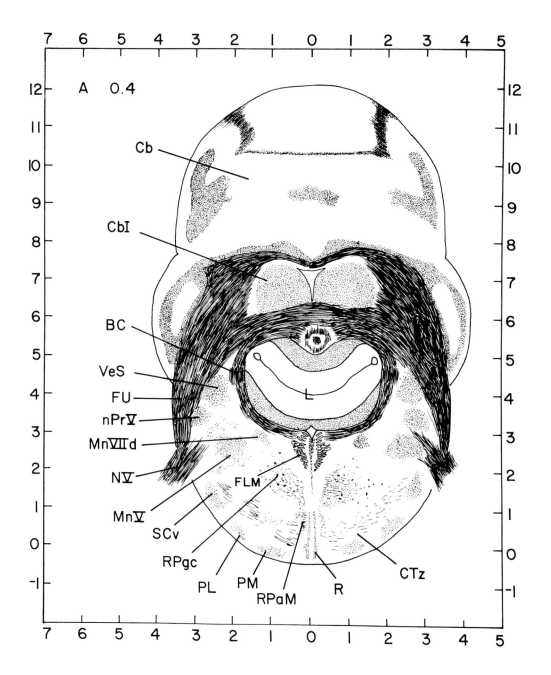

BC Brachium conjunctivum
Cb Cerebellum
CbI Nucleus cerebellaris internus
CTz Corpus trapezoideum (Papez)
FLM Fasciculus longitudinalis medialis
FU Fasciculus uncinatus (Russell)
L Lingula; vinculum lingulae (ICAAN)
Mn V Nucleus motorius nervi trigemini
Mn VII d Nucleus motorius nervi facialis, pars dorsalis
nPr V Nucleus sensorius principalis nervi trigemini
N V Nervus trigeminus
PL Nucleus pontis lateralis
PM Nucleus pontis medialis
R Nucleus raphes (Raphe nucleus)
RPaM Nucleus reticularis paramedianus (ICAAN);
nucleus paramedianus (Karten and Hodos)
RPgc Nucleus reticularis pontis caudalis, pars
gigantocellularis
SCv Nucleus subceruleus ventralis
VeS Nucleus vestibularis superior

BC   Brachium conjunctivum
Cb   Cerebellum
CbI   Nucleus cerebellaris internus
CbIM   Nucleus cerebellaris intermedius
CCV   Commissura cerebellaris ventralis
CTz   Corpus trapezoideum (Papez)
FLM   Fasciculus longitudinalis medialis
FU   Fasciculus uncinatus (Russell)
FUm   Fasciculus uncinatus (Russell), pars medialis
L   Lingula; vinculum lingulae (ICAAN)
Mn V   Nucleus motorius nervi trigemini
Mn VII d   Nucleus motorius nervi facialis, pars dorsalis
Mn VII v   Nucleus motorius nervi facialis, pars ventralis
N V   Nervus trigeminus
N VII   Nervus facialis
PL   Nucleus pontis lateralis
PM   Nucleus pontis medialis
R   Nucleus raphes (Raphe nucleus)
RPaM   Nucleus reticularis paramedianus (ICAAN); nucleus paramedianus (Karten and Hodos)
RPgc   Nucleus reticularis pontis caudalis, pars gigantocellularis
SCv   Nucleus subceruleus ventralis

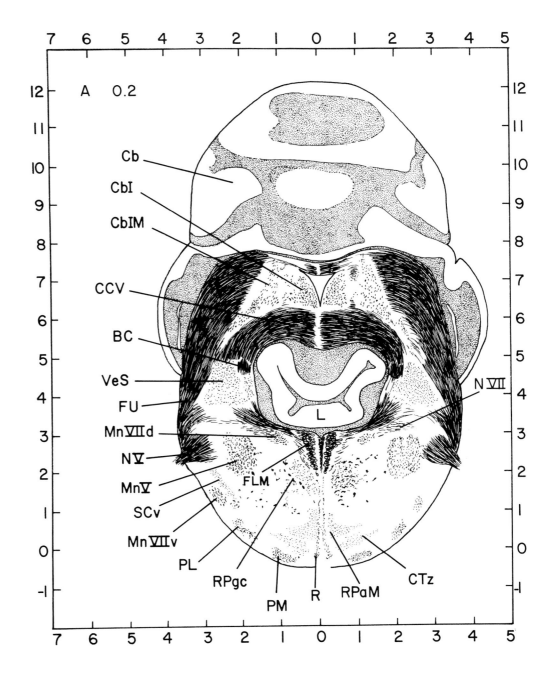

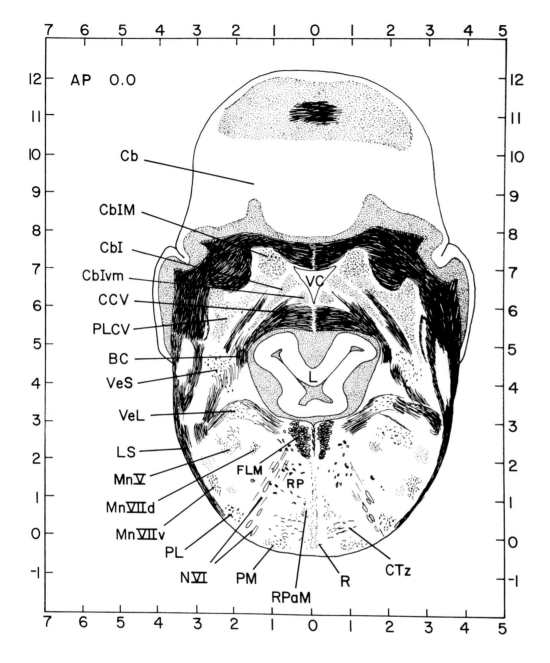

BC　Brachium conjunctivum
Cb　Cerebellum
CbI　Nucleus cerebellaris internus
CbIM　Nucleus cerebellaris intermedius
CbIvm　Nucleus cerebellaris internus, pars
　　　ventromedialis
CCV　Commissura cerebellaris ventralis
CTz　Corpus trapezoideum (Papez)
FLM　Fasciculus longitudinalis medialis
L　Lingula; vinculum lingulae (ICAAN)
LS　Lemniscus spinalis
Mn V　Nucleus motorius nervi trigemini
Mn VII d　Nucleus motorius nervi facialis, pars dorsalis
Mn VII v　Nucleus motorius nervi facialis, pars
　　　ventralis
N VI　Nervus abducens
PL　Nucleus pontis lateralis
PLCV　Processus lateralis cerebello-vestibularis
PM　Nucleus pontis medialis
R　Nucleus raphes (Raphe nucleus)
RP　Nucleus reticularis pontis caudalis
RPaM　Nucleus reticularis paramedianus (ICAAN);
　　　nucleus paramedianus (Karten and Hodos)
VC　Ventriculus cerebelli
VeL　Nucleus vestibularis lateralis
VeS　Nucleus vestibularis superior

BC   Brachium conjunctivum
Cb   Cerebellum
CbI   Nucleus cerebellaris internus
CbIM   Nucleus cerebellaris intermedius
CbIvm   Nucleus cerebellaris internus, pars
            ventromedialis
CbL   Nucleus cerebellaris lateralis
CCV   Commissura cerebellaris ventralis
CTz   Corpus trapezoideum (Papez)
FLM   Fasciculus longitudinalis medialis
L   Lingula; vinculum lingulae (ICAAN)
LO   Tractus lamino-olivaris
LS   Lemniscus spinalis
Mn VII d   Nucleus motorius nervi facialis, pars dorsalis
Mn VII v   Nucleus motorius nervi facialis, pars
            ventralis
n VI   Nucleus nervi abducentis
N VI   Nervus abducens
OS   Nucleus olivaris superior
PL   Nucleus pontis lateralis
PM   Nucleus pontis medialis
R   Nucleus raphes (Raphe nucleus)
RP   Nucleus reticularis pontis caudalis
RPaM   Nucleus reticularis paramedianus (ICAAN);
            nucleus paramedianus (Karten and Hodos)
VC   Ventriculus cerebelli
VeL   Nucleus vestibularis lateralis
VeM   Nucleus vestibularis medialis
VeS   Nucleus vestibularis superior

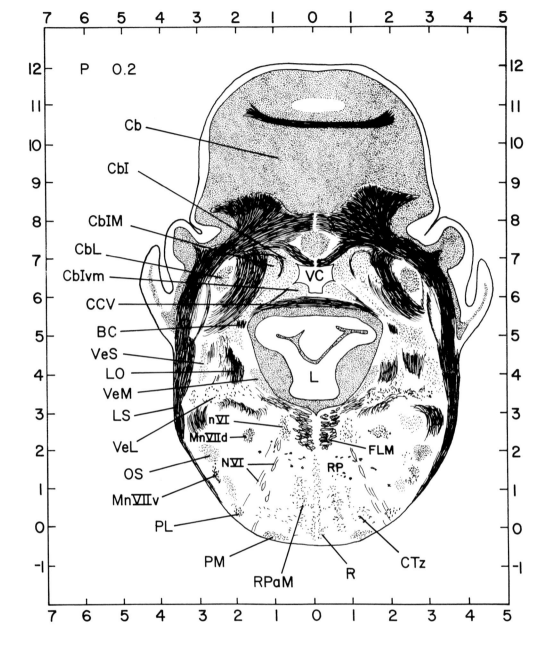

P 0.2

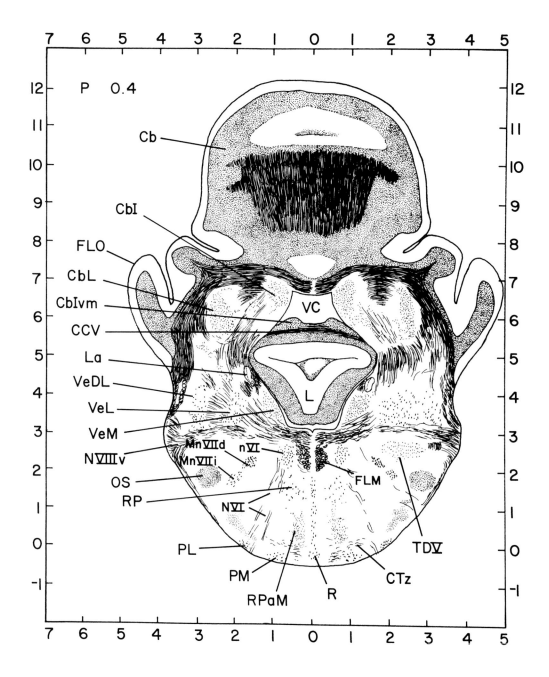

Cb   Cerebellum
CbI   Nucleus cerebellaris internus
CBIvm   Nucleus cerebellaris internus, pars
        ventromedialis
CbL   Nucleus cerebellaris lateralis
CCV   Commissura cerebellaris ventralis
CTz   Corpus trapezoideum (Papez)
FLM   Fasciculus longitudinalis medialis
FLO   Flocculus
L   Lingula; vinculum lingulae (ICAAN)
La   Nucleus laminaris
Mn VII d   Nucleus motorius nervi facialis, pars dorsalis
Mn VII i   Nucleus motorius nervi facialis, pars
        intermedia
n VI   Nucleus nervi abducentis
N VI   Nervus abducens
N VIII v   Nervus octavus, pars vestibularis
OS   Nucleus olivaris superior
PL   Nucleus pontis lateralis
PM   Nucleus pontis medialis
R   Nucleus raphes (Raphe nucleus)
RP   Nucleus reticularis pontis caudalis
RPaM   Nucleus reticularis paramedianus (ICAAN);
        nucleus paramedianus (Karten and Hodos)
TD V   Nucleus et tractus descendens nervi
        trigemini
VC   Ventriculus cerebelli
VeDL   Nucleus vestibularis dorsolateralis (Sanders)
VeL   Nucleus vestibularis lateralis
VeM   Nucleus vestibularis medialis

Cb   Cerebellum
CbL   Nucleus cerebellaris lateralis
CCV   Commissura cerebellaris ventralis
CTz   Corpus trapezoideum (Papez)
FLM   Fasciculus longitudinalis medialis
FLO   Flocculus
L   Lingula; vinculum lingulae (ICAAN)
La   Nucleus laminaris
LS   Lemniscus spinalis
n VI Nucleus nervi abducentis
N VI Nervus abducens
N VIII v Nervus octavus, pars vestibularis
OS   Nucleus olivaris superior
PL   Nucleus pontis lateralis
PM   Nucleus pontis medialis
R   Nucleus raphes (Raphe nucleus)
RP   Nucleus reticularis pontis caudalis
RPaM Nucleus reticularis paramedianus (ICAAN);
      nucleus paramedianus (Karten and Hodos)
Ta   Nucleus tangentialis (Cajal)
TD V Nucleus et tractus descendens nervi
      trigemini
VeDL Nucleus vestibularis dorsolateralis (Sanders)
VeL   Nucleus vestibularis lateralis
VeM   Nucleus vestibularis medialis

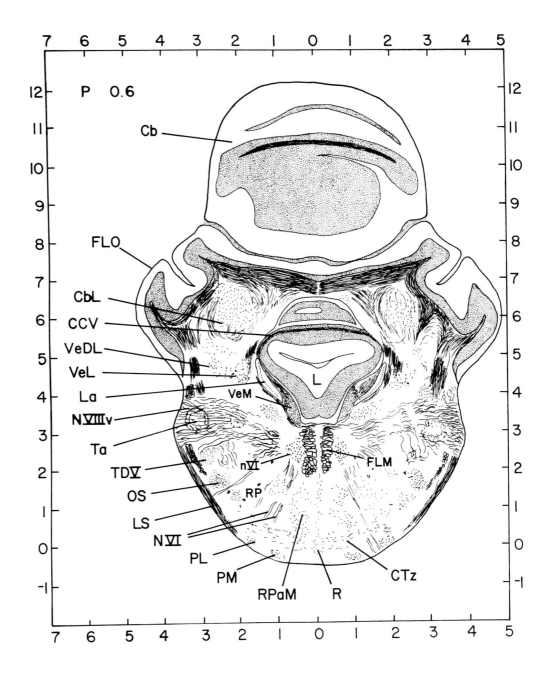

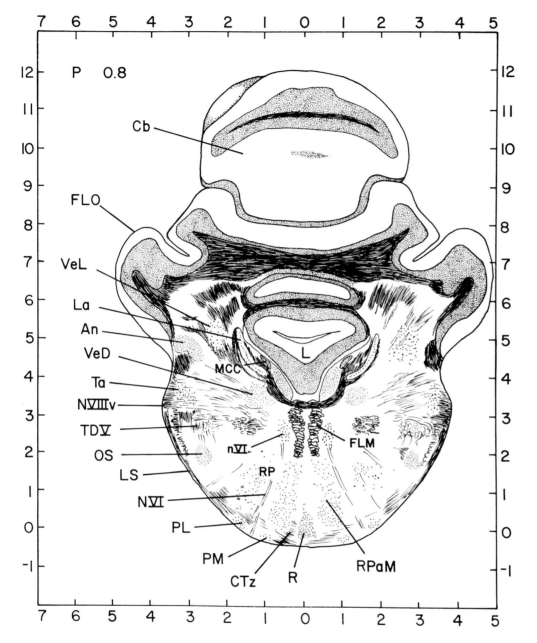

An   Nucleus angularis
Cb   Cerebellum
CTz   Corpus trapezoideum (Papez)
FLM   Fasciculus longitudinalis medialis
FLO   Flocculus
L   Lingula; vinculum lingulae (ICAAN)
La   Nucleus laminaris
LS   Lemniscus spinalis
MCC   Nucleus magnocellularis cochlearis
n VI Nucleus nervi abducentis
N VI Nervus abducens
N VIII v Nervus octavus, pars vestibularis
OS   Nucleus olivaris superior
PL   Nucleus pontis lateralis
PM   Nucleus pontis medialis
R   Nucleus raphes (Raphe nucleus)
RP   Nucleus reticularis pontis caudalis
RPaM Nucleus reticularis paramedianus (ICAAN);
      nucleus paramedianus (Karten and Hodos)
Ta   Nucleus tangentialis (Cajal)
TD V Nucleus et tractus descendens nervi trigemini
VeL   Nucleus vestibularis lateralis
VeD   Nucleus vestibularis descendens

An   Nucleus angularis
Cb   Cerebellum
CTz   Corpus trapezoideum (Papez)
FLM   Fasciculus longitudinalis medialis
L   Lingula; vinculum lingulae (ICAAN)
La   Nucleus laminaris
LS   Lemniscus spinalis
MCC   Nucleus magnocellularis cochlearis
n VI   Nucleus nervi abducentis
N VI   Nervus abducens
N VIII v   Nervus octavus, pars vestibularis
OS   Nucleus olivaris superior
R   Nucleus raphes (Raphe nucleus)
Rgc   Nucleus reticularis gigantocellularis
RP   Nucleus reticularis pontis caudalis
RPaM   Nucleus reticularis paramedianus (ICAAN);
         nucleus paramedianus (Karten and Hodos)
Ta   Nucleus tangentialis (Cajal)
TD V   Nucleus et tractus descendens nervi trigemini
VeD   Nucleus vestibularis descendens
VeL   Nucleus vestibularis lateralis
VeM   Nucleus vestibularis medialis
V IV   Ventriculus quartus (Fourth ventricle)

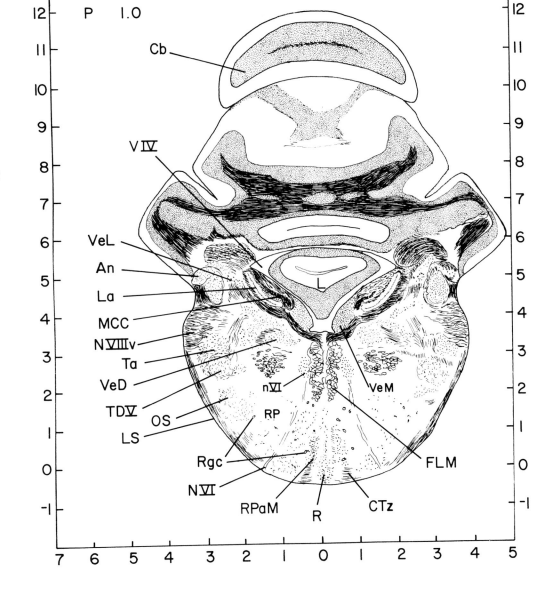

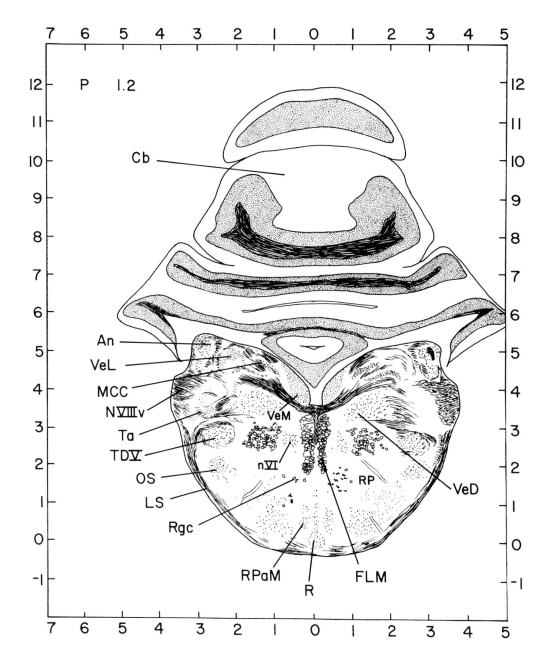

An   Nucleus angularis
Cb   Cerebellum
FLM   Fasciculus longitudinalis medialis
LS   Lemniscus spinalis
MCC   Nucleus magnocellularis cochlearis
n VI   Nucleus nervi abducentis
N VIII v   Nervus octavus, pars vestibularis
OS   Nucleus olivaris superior
R   Nucleus raphes (Raphe nucleus)
Rgc   Nucleus reticularis gigantocellularis
RP   Nucleus reticularis pontis caudalis
RPaM   Nucleus reticularis paramedianus (ICAAN);
nucleus paramedianus (Karten and Hodos)
Ta   Nucleus tangentialis (Cajal)
TD V   Nucleus et tractus descendens nervi trigemini
VeD   Nucleus vestibularis descendens
VeL   Nucleus vestibularis lateralis
VeM   Nucleus vestibularis medialis

An   Nucleus angularis
Cb   Cerebellum
FLM   Fasciculus longitudinalis medialis
LS   Lemniscus spinalis
MCC   Nucleus magnocellularis cochlearis
N VIII c   Nervus octavus, pars cochlearis
R   Nucleus raphes (Raphe nucleus)
Rgc   Nucleus reticularis gigantocellularis
RPaM   Nucleus reticularis paramedianus (ICAAN);
         nucleus paramedianus (Karten and Hodos)
Rpc   Nucleus reticularis parvocellularis
SCbd   Tractus spinocerebellaris dorsalis
TD V   Nucleus et tractus descendens nervi trigemini
VeD   Nucleus vestibularis descendens
VeM   Nucleus vestibularis medialis

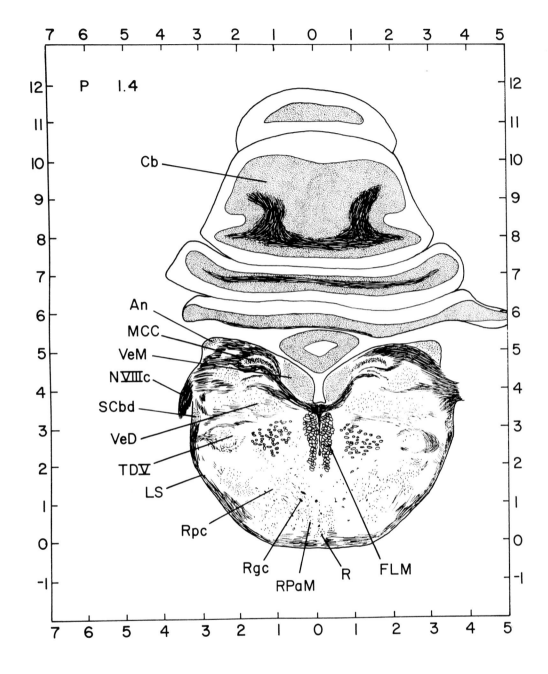

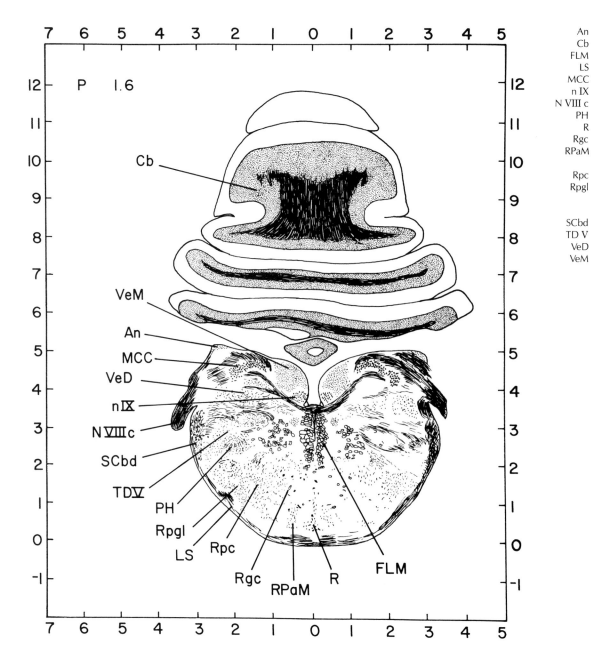

7 6 5 4 3 2 1 0 1 2 3 4 5

P   1.6

12   12

11   11

10   10

Cb

9   9

8   8

7   7

VeM

6   6

An

5   5

MCC

VeD

4   4

nIX

3   3

NVIIIc

SCbd

2   2

TDV

PH

1   1

Rpgl

Rpc

0   R   FLM   0

LS

Rgc RPaM

-1   -1

7 6 5 4 3 2 1 0 1 2 3 4 5

An   Nucleus angularis
Cb   Cerebellum
FLM Fasciculus longitudinalis medialis
LS   Lemniscus spinalis
MCC Nucleus magnocellularis cochlearis
n IX Nucleus nervi glossopharyngei
N VIII c Nervus octavus, pars cochlearis
PH   Plexus of Horsley
R   Nucleus raphes (Raphe nucleus)
Rgc Nucleus reticularis gigantocellularis
RPaM Nucleus reticularis paramedianus (ICAAN);
      nucleus paramedianus (Karten and Hodos)
Rpc Nucleus reticularis parvocellularis
Rpgl Nucleus reticularis paragigantocellularis
      lateralis (ICAAN); nucleus paragigan-
      tocellularis lateralis (Karten and Hodos)
SCbd Tractus spinocerebellaris dorsalis
TD V Nucleus et tractus descendens nervi trigemini
VeD Nucleus vestibularis descendens
VeM Nucleus vestibularis medialis

Cb   Cerebellum
FLM   Fasciculus longitudinalis medialis
LM   Lemniscus medialis
LS   Lemniscus spinalis
MCC   Nucleus magnocellularis cochlearis
n IX Nucleus nervi glossopharyngei
N VIII c Nervus octavus, pars cochlearis
PH   Plexus of Horsley
R   Nucleus raphes (Raphe nucleus)
Rgc   Nucleus reticularis gigantocellularis
RPaM Nucleus reticularis paramedianus (ICAAN);
      nucleus paramedianus (Karten and Hodos)
Rpc   Nucleus reticularis parvocellularis
Rpgl Nucleus reticularis paragigantocellularis
      lateralis (ICAAN); nucleus paragigan-
      tocellularis lateralis (Karten and Hodos)
RST   Nucleus reticularis subtrigeminalis
SCbd Tractus spinocerebellaris dorsalis
TD V Nucleus et tractus descendens nervi trigemini
VeD   Nucleus vestibularis descendens
VeM   Nucleus vestibularis medialis

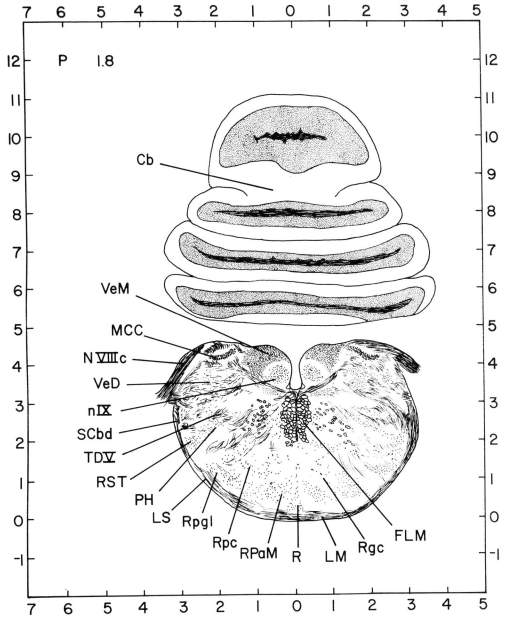

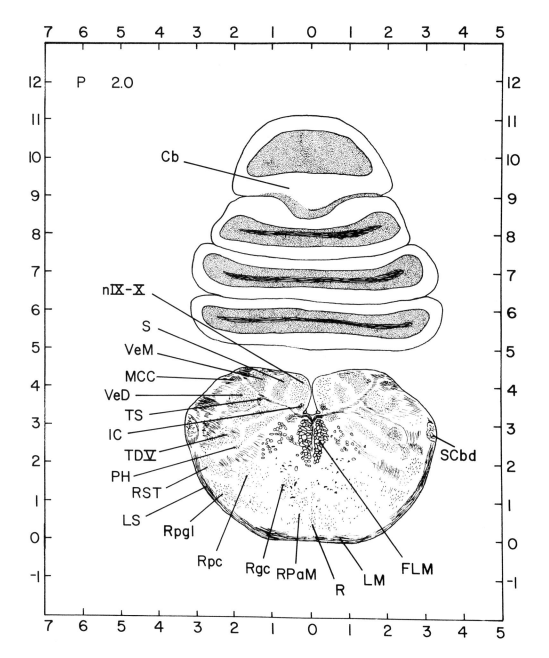

P 2.0

Cb — Cerebellum
FLM — Fasciculus longitudinalis medialis
IC — Nucleus intercalatus
LM — Lemniscus medialis
LS — Lemniscus spinalis
MCC — Nucleus magnocellularis cochlearis
n IX-X — Nucleus nervi glossopharyngei et nucleus motorius dorsalis nervi vagi
PH — Plexus of Horsley
R — Nucleus raphes (Raphe nucleus)
Rgc — Nucleus reticularis gigantocellularis
RPaM — Nucleus reticularis paramedianus (ICAAN); nucleus paramedianus (Karten and Hodos)
Rpc — Nucleus reticularis parvocellularis
Rpgl — Nucleus reticularis paragigantocellularis lateralis (ICAAN); nucleus paragigantocellularis lateralis (Karten and Hodos)
RST — Nucleus reticularis subtrigeminalis
S — Nucleus tractus solitarii
SCbd — Tractus spinocerebellaris dorsalis
TS — Tractus solitarius
TD V — Nucleus et tractus descendens nervi trigemini
VeD — Nucleus vestibularis descendens
VeM — Nucleus vestibularis medialis

Cb   Cerebellum
FLM   Fasciculus longitudinalis medialis
IC   Nucleus intercalatus
LM   Lemniscus medialis
LS   Lemniscus spinalis
n IX-X   Nucleus nervi glossopharyngei et nucleus
         motorius dorsalis nervi vagi
PH   Plexus of Horsley
R   Nucleus raphes (Raphe nucleus)
Rgc   Nucleus reticularis gigantocellularis
RPaM   Nucleus reticularis paramedianus (ICAAN);
         nucleus paramedianus (Karten and Hodos)
Rpc   Nucleus reticularis parvocellularis
Rpgl   Nucleus reticularis paragigantocellularis
         lateralis (ICAAN); nucleus paragigantocellularis
         lateralis (Karten and Hodos)
RST   Nucleus reticularis subtrigeminalis
S   Nucleus tractus solitarii
TS   Tractus solitarius
TD V   Nucleus et tractus descendens nervi trigemini
VeD   Nucleus vestibularis descendens
VeM   Nucleus vestibularis medialis

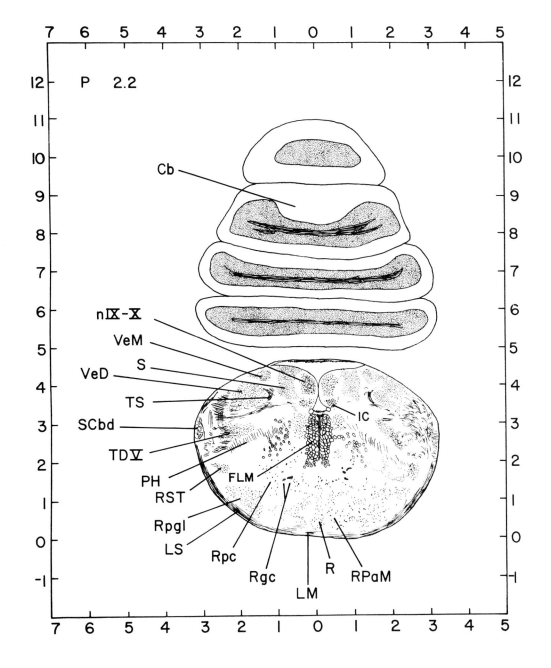

P 2.2

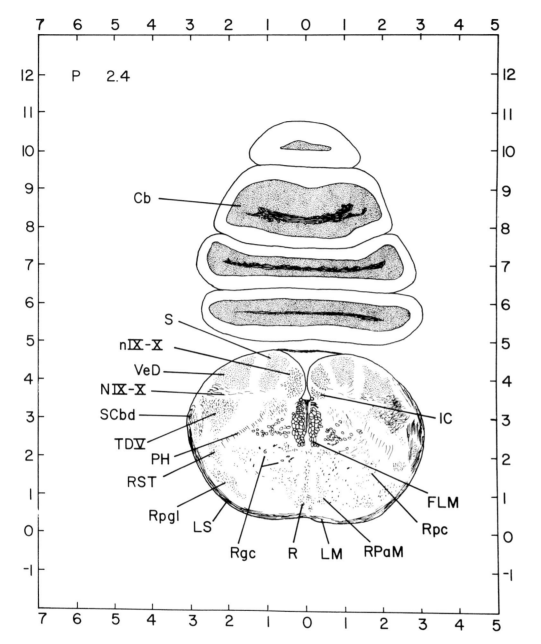

Cb Cerebellum
FLM Fasciculus longitudinalis medialis
IC Nucleus intercalatus
LM Lemniscus medialis
LS Lemniscus spinalis
n IX-X Nucleus nervi glossopharyngei et nucleus
       motorius dorsalis nervi vagi
N IX-X Nervi glossopharyngeus et vagus
PH Plexus of Horsley
R Nucleus raphes (Raphe nucleus)
Rgc Nucleus reticularis gigantocellularis
RPaM Nucleus reticularis paramedianus (ICAAN);
       nucleus paramedianus (Karten and Hodos)
Rpc Nucleus reticularis parvocellularis
Rpgl Nucleus reticularis paragigantocellularis
       lateralis (ICAAN); nucleus paragigan-
       tocellularis lateralis (Karten and Hodos)
RST Nucleus reticularis subtrigeminalis
S Nucleus tractus solitarii
SCbd Tractus spinocerebellaris dorsalis
TD V Nucleus et tractus descendens nervi trigemini
VeD Nucleus vestibularis descendens

APa Area postrema
Cb   Cerebellum
CE   Nucleus cuneatus externus (Karten and
     Hodos); nucleus cuneatus accessorius
     [lateralis] (ICAAN)
FLM Fasciculus longitudinalis medialis
IC   Nucleus intercalatus
LM   Lemniscus medialis
LS   Lemniscus spinalis
n IX-X Nucleus nervi glossopharyngei et nucleus
        motorius dorsalis nervi vagi
N IX-X Nervi glossopharyngeus et vagus
n XII   Nucleus nervi hypoglossi (Nottebohm, Stokes,
        and Leonard), pars tracheosyringealis, pars
        lingualis; nucleus nervi cervicalis medialis
        (Watanabe, Iwata, and Yasuda)
PH   Plexus of Horsley
R   Nucleus raphes (Raphe nucleus)
Rgc Nucleus reticularis gigantocellularis
Rpc Nucleus reticularis parvocellularis
Rpgl Nucleus reticularis paragigantocellularis
     lateralis (ICAAN); nucleus paragigan-
     tocellularis lateralis (Karten and Hodos)
RST Nucleus reticularis subtrigeminalis
S   Nucleus tractus solitarii
TD V Nucleus et tractus descendens nervi trigemini
VeD Nucleus vestibularis descendens

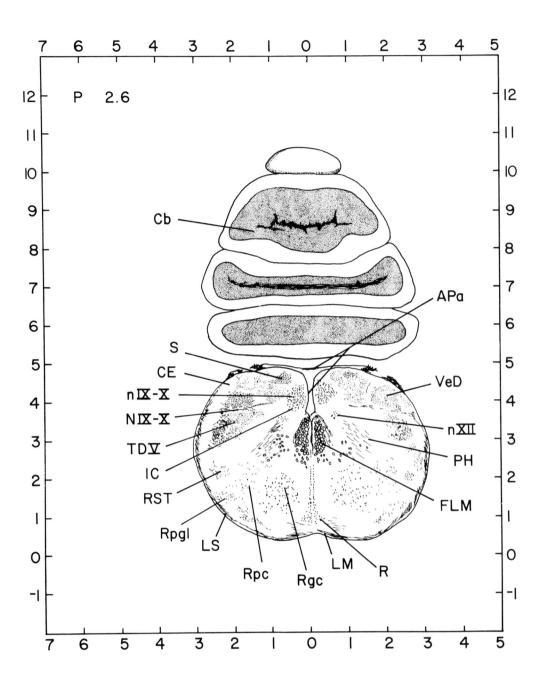

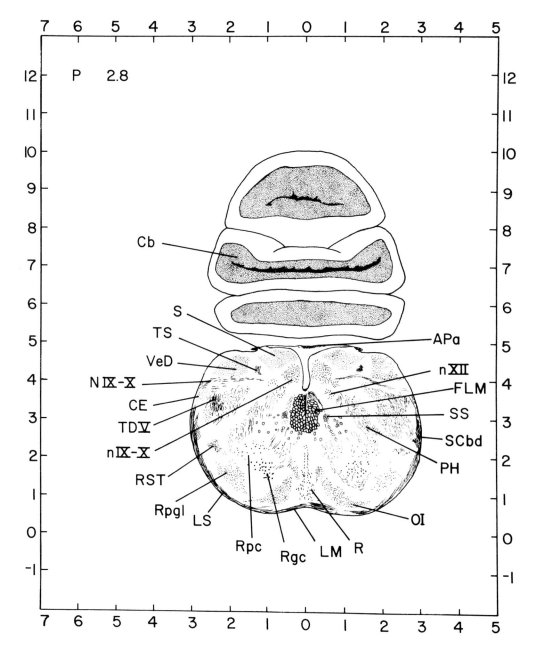

Cb Cerebellum
CE Nucleus cuneatus externus (Karten and
      Hodos); nucleus cuneatus accessorius
      [lateralis] (ICAAN)
FLM Fasciculus longitudinalis medialis
LM Lemniscus medialis
LS Lemniscus spinalis
Mn X Nucleus motorius dorsalis nervi vagi
N IX-X Nervi glossopharyngeus et vagus
n XII Nucleus nervi hypoglossi (Nottebohm, Stokes,
      and Leonard), pars tracheosyringealis, pars
      lingualis; nucleus nervi cervicalis medialis
      (Watanabe, Iwata, and Yasuda)
OI Nucleus olivaris inferior (Kooy and
      Vogt-Nilsen); complexus olivaris caudalis
      (ICAAN)
PH Plexus of Horsley
R Nucleus raphes (Raphe nucleus)
RL Nucleus reticularis lateralis
RST Nucleus reticularis subtrigeminalis
S Nucleus tractus solitarii
SCbd Tractus spinocerebellaris dorsalis
SS Nucleus supraspinalis (Wild and Zeigler)
TS Tractus solitarius
TD V Nucleus et tractus descendens nervi trigemini
VeD Nucleus vestibularis descendens

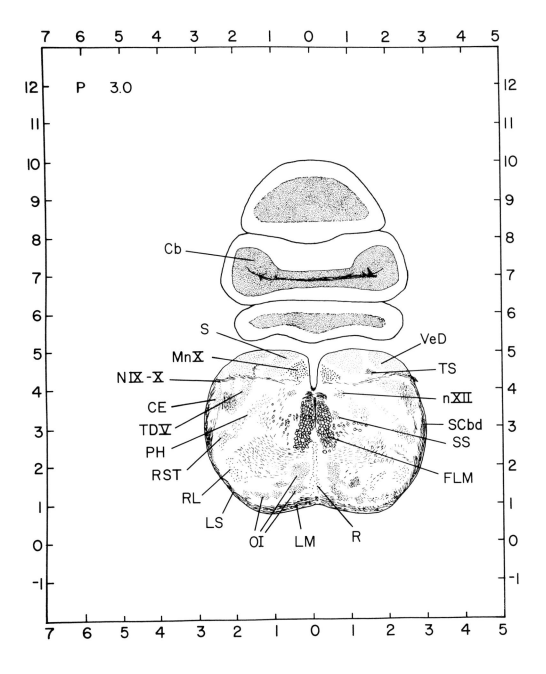

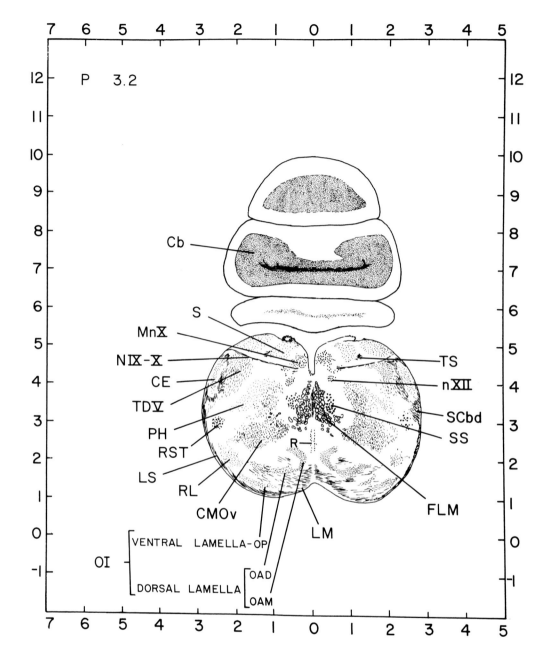

Cb   Cerebellum
CE   Nucleus cuneatus externus (Karten and Hodos); nucleus cuneatus accessorius [lateralis] (ICAAN)
CMOv Nucleus centralis medullae oblongatae, pars ventralis
FLM   Fasciculus longitudinalis medialis
LM   Lemniscus medialis
LS   Lemniscus spinalis
Mn X Nucleus motorius dorsalis nervi vagi
N IX-X Nervi glossopharyngeus et vagus
n XII Nucleus nervi hypoglossi (Nottebohm, Stokes, and Leonard), pars tracheosyringealis, pars lingualis; nucleus nervi cervicalis medialis (Watanabe, Iwata, and Yasuda)
OI   Nucleus olivaris inferior (Kooy and Vogt-Nilsen); complexus olivaris caudalis (ICAAN); components include:
      OAD Nucleus olivaris accessorius dorsalis
      OAM Nucleus olivaris accessorius medialis
      OP   Nucleus olivaris principalis
PH   Plexus of Horsley
R   Nucleus raphes (Raphe nucleus)
RL   Nucleus reticularis lateralis
RST   Nucleus reticularis subtrigeminalis
S   Nucleus tractus solitarii
SCbd Tractus spinocerebellaris dorsalis
SS   Nucleus supraspinalis (Wild and Zeigler)
TS   Tractus solitarius
TD V Nucleus et tractus descendens nervi trigemini

Cb   Cerebellum
CE   Nucleus cuneatus externus (Karten and
       Hodos); nucleus cuneatus accessorius
       [lateralis] (ICAAN)
CMOd Nucleus centralis medullae oblongatae, pars
       dorsalis
CMOv Nucleus centralis medullae oblongatae, pars
       ventralis
FLM   Fasciculus longitudinalis medialis
LM   Lemniscus medialis
LS   Lemniscus spinalis
Mn X Nucleus motorius dorsalis nervi vagi
n XII Nucleus nervi hypoglossi (Nottebohm, Stokes,
       and Leonard), pars tracheosyringealis, pars
       lingualis; nucleus nervi cervicalis medialis
       (Watanabe, Iwata, and Yasuda)
N X   Nervus vagus
OI   Nucleus olivaris inferior (Kooy and
       Vogt-Nilsen); complexus olivaris caudalis
       (ICAAN)
R   Nucleus raphes (Raphe nucleus)
RL   Nucleus reticularis lateralis
RST Nucleus reticularis subtrigeminalis
S   Nucleus tractus solitarii
SCbd Tractus spinocerebellaris dorsalis
SG   Substantia gelatinosa Rolandi (trigemini)
SS   Nucleus supraspinalis (Wild and Zeigler)
TS   Tractus solitarius
TD V Nucleus et tractus descendens nervi trigemini

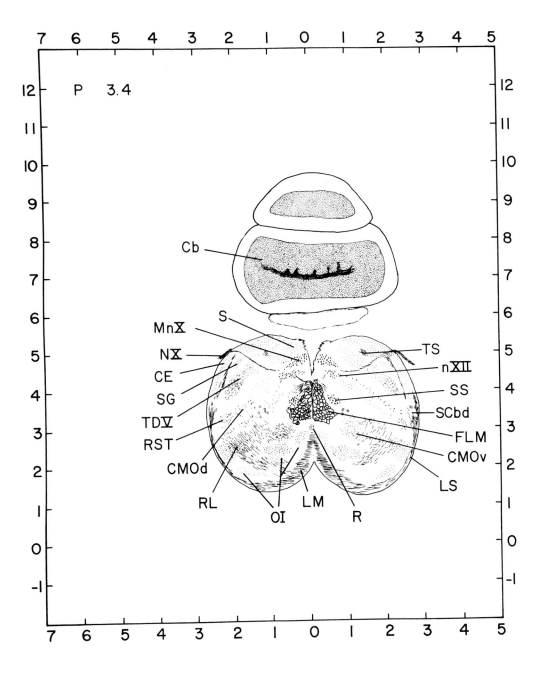

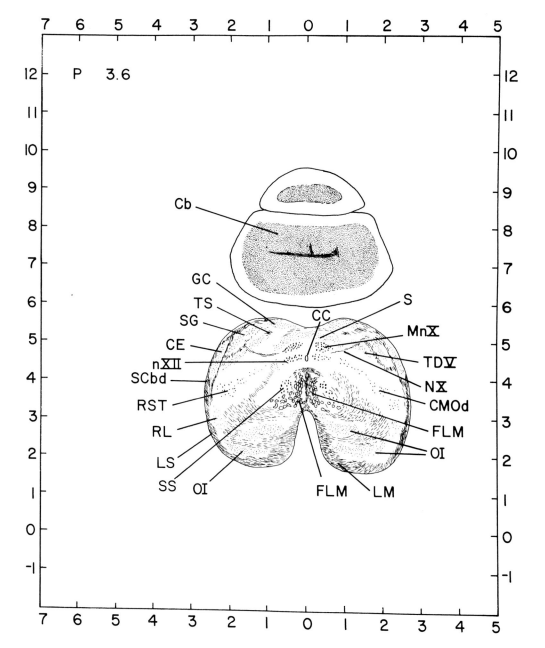

Cb   Cerebellum
CC   Canalis centralis
CMOd   Nucleus centralis medullae oblongatae, pars
       dorsalis
CMOv   Nucleus centralis medullae oblongatae, pars
       ventralis
FD   Funiculus dorsalis
FLM   Fasciculus longitudinalis medialis
GC   Nuclei gracilis et cuneatus
Mn X   Nucleus motorius dorsalis nervi vagi
n XII Nucleus nervi hypoglossi (Nottebohm, Stokes,
       and Leonard), pars tracheosyringealis, pars
       lingualis; nucleus nervi cervicalis medialis
       (Watanabe, Iwata, and Yasuda)
N XII Nervus hypoglossus
OI   Nucleus olivaris inferior (Kooy and
       Vogt-Nilsen); complexus olivaris caudalis
       (ICAAN)
RL   Nucleus reticularis lateralis
S   Nucleus tractus solitarii
SCbd   Tractus spinocerebellaris dorsalis
SG   Substantia gelatinosa Rolandi (Trigemini)
SS   Nucleus supraspinalis (Wild and Zeigler)
TD V   Nucleus et tractus descendens nervi trigemini

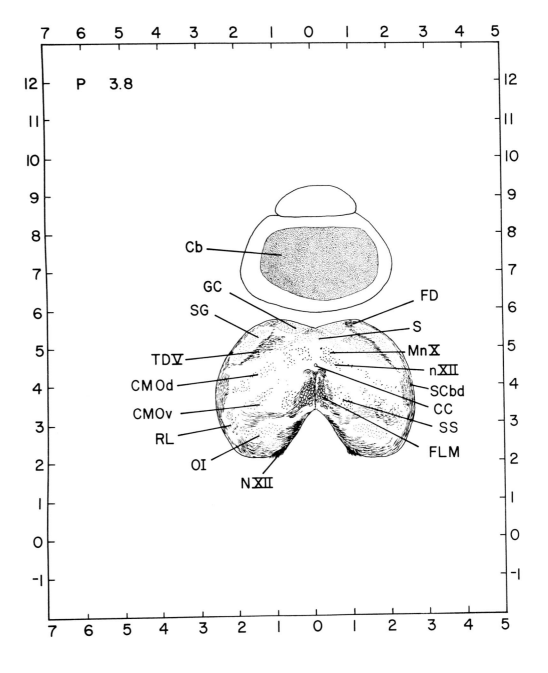

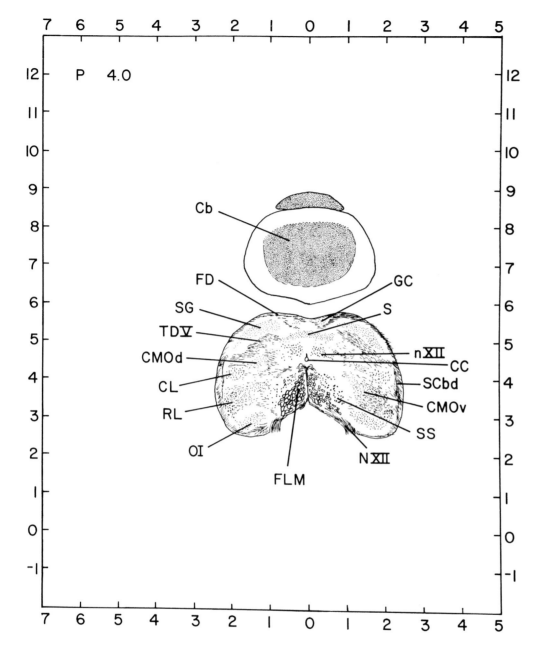

Cb   Cerebellum
CC   Canalis centralis
CL   Nucleus cervicalis lateralis
CMOd Nucleus centralis medullae oblongatae, pars
      dorsalis
CMOv Nucleus centralis medullae oblongatae, pars
      ventralis
FD   Funiculus dorsalis
FLM   Fasciculus longitudinalis medialis
GC   Nuclei gracilis et cuneatus
n XII Nucleus nervi hypoglossi (Nottebohm, Stokes,
      and Leonard), pars tracheosyringealis, pars
      lingualis; nucleus nervi cervicalis medialis
      (Watanabe, Iwata, and Yasuda)
N XII Nervus hypoglossus
OI   Nucleus olivaris inferior (Kooy and
      Vogt-Nilsen); complexus olivaris caudalis
      (ICAAN)
RL   Nucleus reticularis lateralis
S   Nucleus tractus solitarii
SCbd Tractus spinocerebellaris dorsalis
SG   Substantia gelatinosa Rolandi (Trigemini)
SS   Nucleus supraspinalis (Wild and Zeigler)
TD V Nucleus et tractus descendens nervi trigemini

Cb   Cerebellum
CC   Canalis centralis
CL   Nucleus cervicalis lateralis
FD   Funiculus dorsalis
FLt   Funiculus lateralis
FV   Funiculus ventralis
GC   Nuclei gracilis et cuneatus
n XII   Nucleus nervi hypoglossi (Nottebohm, Stokes,
        and Leonard), pars tracheosyringealis, pars
        lingualis; nucleus nervi cervicalis medialis
        (Watanabe, Iwata, and Yasuda)
S   Nucleus tractus solitarii
SG   Substantia gelatinosa Rolandi (Trigemini)
SS   Nucleus supraspinalis (Wild and Zeigler)
TD V   Nucleus et tractus descendens nervi trigemini

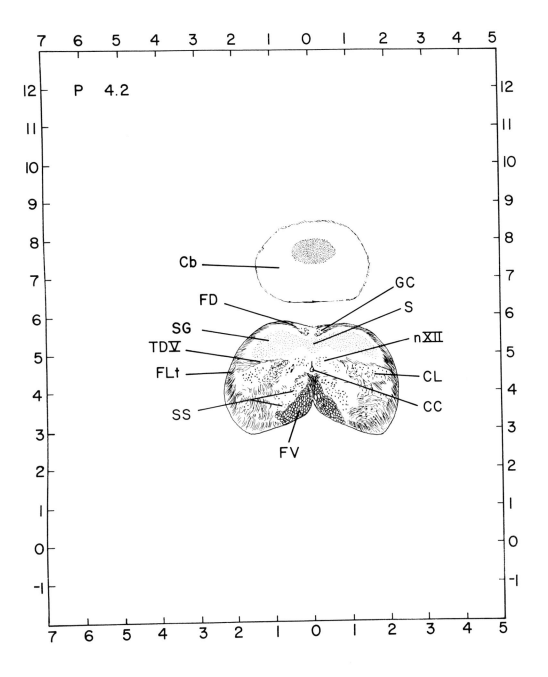

P   4.2

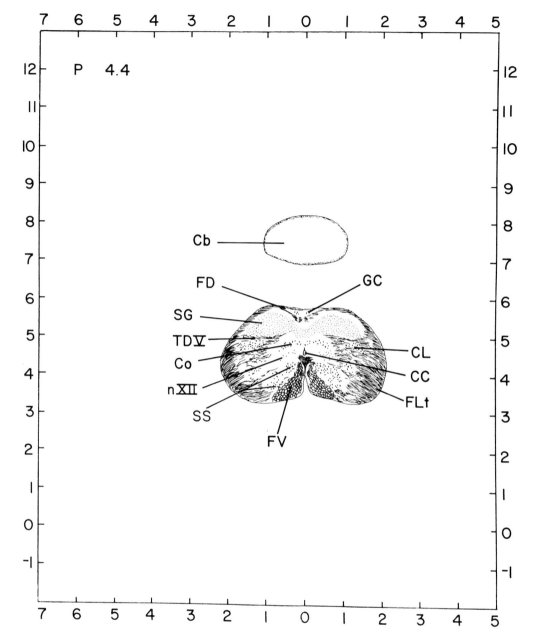

Cb Cerebellum
CC Canalis centralis
CL Nucleus cervicalis lateralis
Co Nucleus commissuralis (Haller)
FD Funiculus dorsalis
FLt Funiculus lateralis
FV Funiculus ventralis
GC Nuclei gracilis et cuneatus
n XII Nucleus nervi hypoglossi (Nottebohm, Stokes, and Leonard), pars tracheosyringealis, pars lingualis; nucleus nervi cervicalis medialis (Watanabe, Iwata, and Yasuda)
SG Substantia gelatinosa Rolandi (Trigemini)
SS Nucleus supraspinalis (Wild and Zeigler)
TD V Nucleus et tractus descendens nervi trigemini

Cb   Cerebellum
CC   Canalis centralis
CL   Nucleus cervicalis lateralis
Co   Nucleus commissuralis (Haller)
FD   Funiculus dorsalis
FLt Funiculus lateralis
FV   Funiculus ventralis
GC   Nuclei gracilis et cuneatus
n XI Nucleus nervi accessorii (Spinal accessory
      nerve [Eden and Correia])
SG   Substantia gelatinosa Rolandi (Trigemini)
SS   Nucleus supraspinalis (Wild and Zeigler)
TD V Nucleus et tractus descendens nervi trigemini

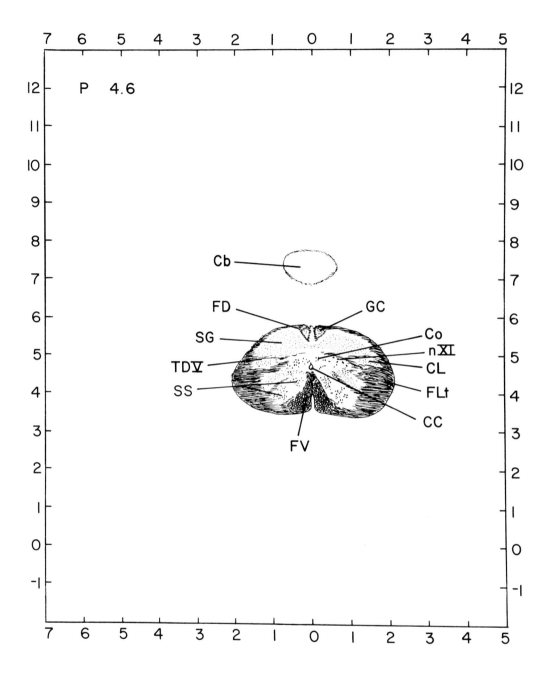

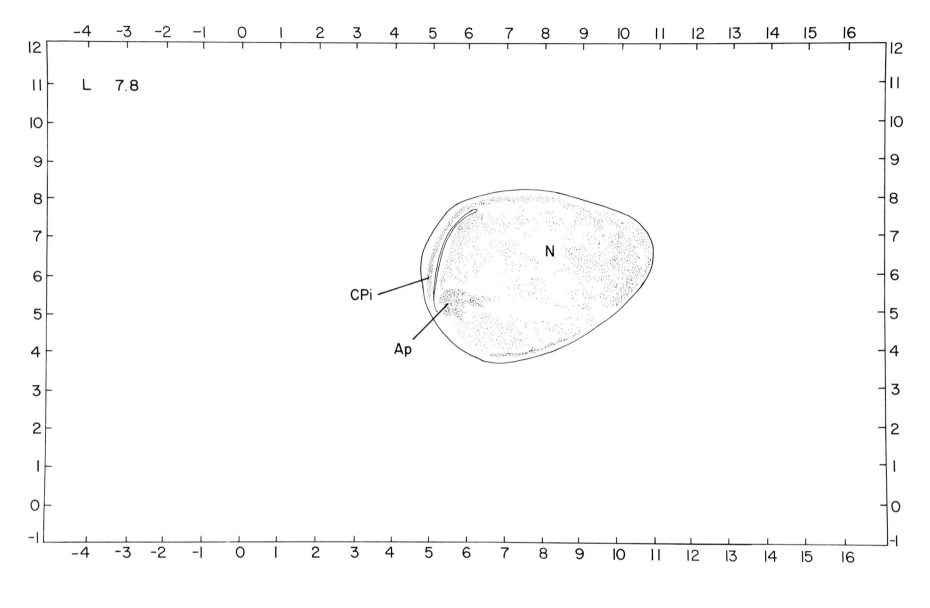

Ap   Archistriatum posterior [caudale] (Zeier and
      Karten)
CPi Cortex piriformis
 N   Neostriatum

PLATE L 7.0   129

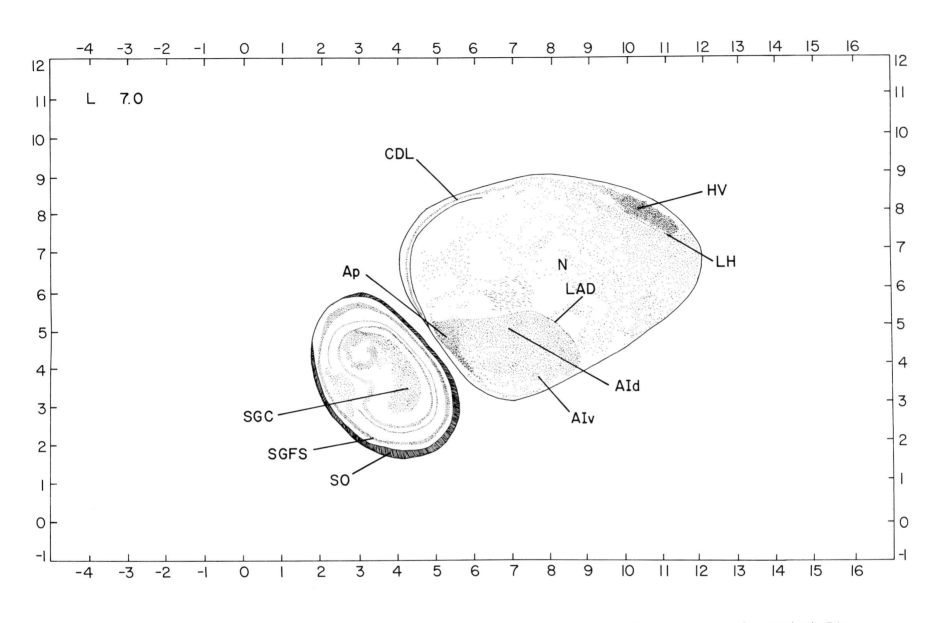

AId Archistriatum intermedium, pars dorsalis (Zeier and Karten)
AIv Archistriatum intermedium, pars ventralis (Zeier and Karten)
Ap Archistriatum posterior [caudale] (Zeier and Karten)
CDL Area corticoidea dorsolateralis
HV Hyperstriatum ventrale
LAD Lamina archistriatalis dorsalis (Zeier and Karten)
LH Lamina hyperstriatica
N Neostriatum
SGC Stratum griseum centrale
SGFS Stratum griseum et fibrosum superficiale
SO Stratum opticum

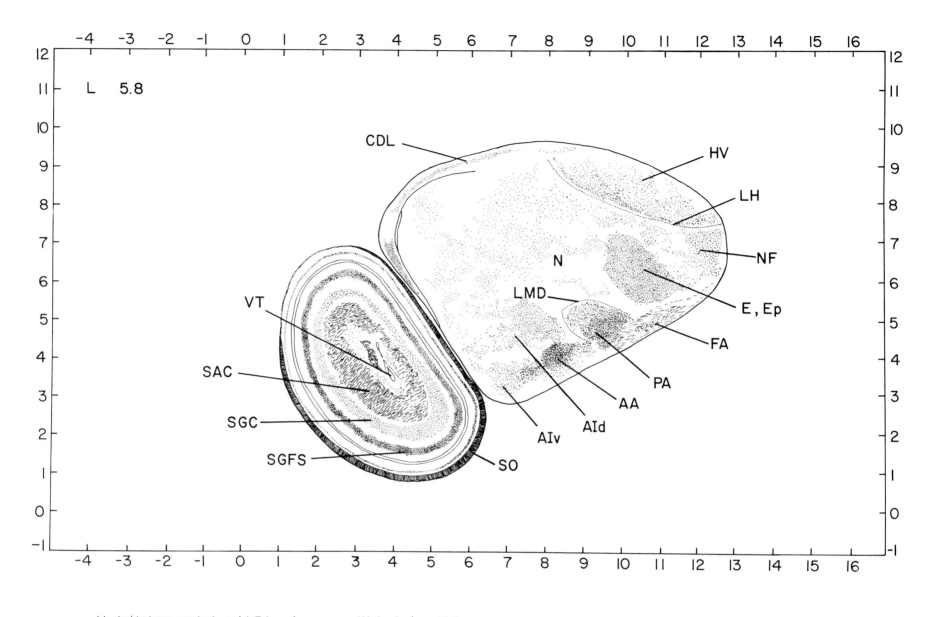

L   5.8

AA   Archistriatum anterior [rostrale] (Zeier and
     Karten)
AId Archistriatum intermedium, pars dorsalis (Zeier
     and Karten)
AIv Archistriatum intermedium, pars ventralis (Zeier
     and Karten)
CDL Area corticoidea dorsolateralis
E, EP Ectostriatum, Cingulum periectostriatale
     (Periectostriatal belt)
FA   Tractus fronto-archistriaticus
HV   Hyperstriatum ventrale

LH   Lamina hyperstriatica
LMD Lamina medullaris dorsalis
N   Neostriatum
NF   Neostriatum frontale
PA   Paleostriatum augmentatum (Caudate putamen)
SAC Stratum album centrale
SGC Stratum griseum centrale
SGFS Stratum griseum et fibrosum superficiale
SO   Stratum opticum
VT   Ventriculus tecti mesencephali

PLATE L 5.0   131

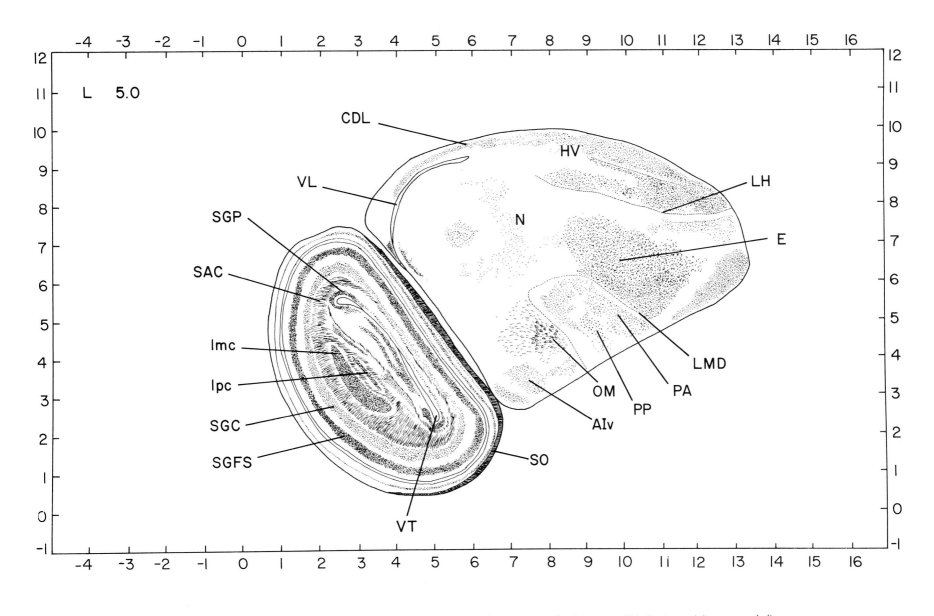

Alv Archistriatum intermedium, pars ventralis (Zeier and Karten)
CDL Area corticoidea dorsolateralis
E Ectostriatum
HV Hyperstriatum ventrale
Imc Nucleus isthmi, pars magnocellularis
Ipc Nucleus isthmi, pars parvocellularis
LH Lamina hyperstriatica
LMD Lamina medullaris dorsalis
N Neostriatum

OM Tractus occipitomesencephalicus
PA Paleostriatum augmentatum (Caudate putamen)
PP Paleostriatum primitivum (Globus pallidus)
SAC Stratum album centrale
SGC Stratum griseum centrale
SGFS Stratum griseum et fibrosum superficiale
SGP Stratum griseum periventriculare
SO Stratum opticum
VL Ventriculus lateralis
VT Ventriculus tecti mesencephali

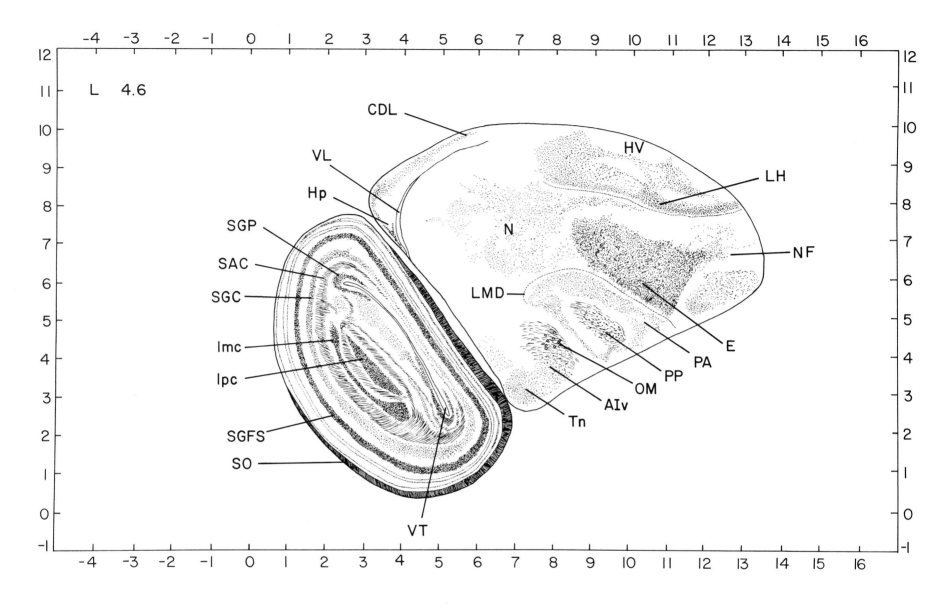

L 4.6

Alv Archistriatum intermedium, pars ventralis (Zeier and Karten)
CDL Area corticoidea dorsolateralis
E Ectostriatum
Hp Hippocampus
HV Hyperstriatum ventrale
Imc Nucleus isthmi, pars magnocellularis
Ipc Nucleus isthmi, pars parvocellularis
LH Lamina hyperstriatica
LMD Lamina medullaris dorsalis
N Neostriatum
NF Neostriatum frontale
OM Tractus occipitomesencephalicus
PA Paleostriatum augmentatum (Caudate putamen)
PP Paleostriatum primitivum (Globus pallidus)
SAC Stratum album centrale
SGC Stratum griseum centrale
SGFS Stratum griseum et fibrosum superficiale
SGP Stratum griseum periventriculare
SO Stratum opticum
Tn Nucleus taeniae
VL Ventriculus lateralis
VT Ventriculus tecti mesencephali

PLATE L 4.2   133

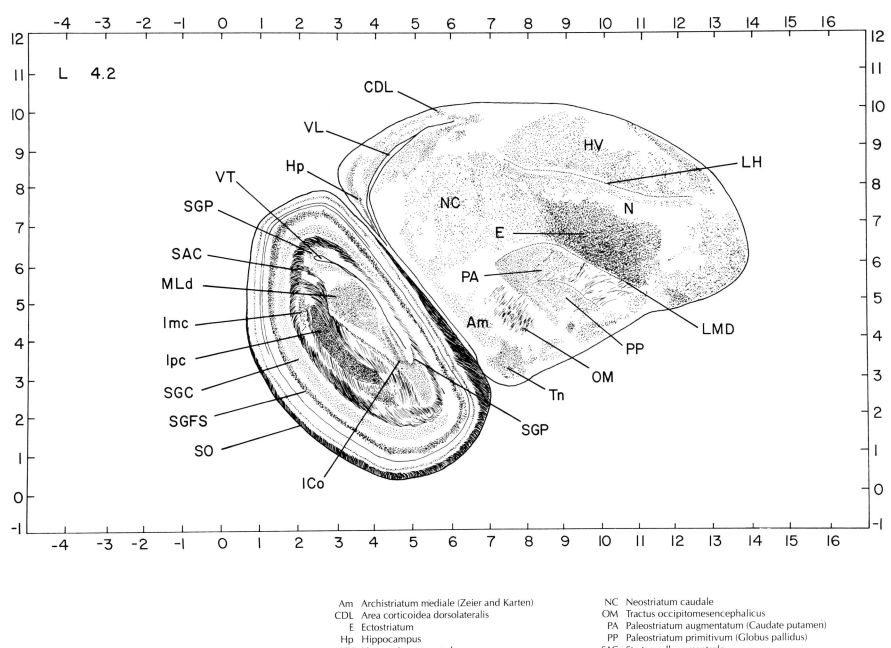

Am	Archistriatum mediale (Zeier and Karten)
CDL	Area corticoidea dorsolateralis
E	Ectostriatum
Hp	Hippocampus
HV	Hyperstriatum ventrale
ICo	Nucleus intercollicularis
Imc	Nucleus isthmi, pars magnocellularis
Ipc	Nucleus isthmi, pars parvocellularis
LH	Lamina hyperstriatica
LMD	Lamina medullaris dorsalis
MLd	Nucleus mesencephalicus lateralis, pars dorsalis
N	Neostriatum

NC	Neostriatum caudale
OM	Tractus occipitomesencephalicus
PA	Paleostriatum augmentatum (Caudate putamen)
PP	Paleostriatum primitivum (Globus pallidus)
SAC	Stratum album centrale
SGC	Stratum griseum centrale
SGFS	Stratum griseum et fibrosum superficiale
SGP	Stratum griseum periventriculare
SO	Stratum opticum
Tn	Nucleus taeniae
VL	Ventriculus lateralis
VT	Ventriculus tecti mesencephali

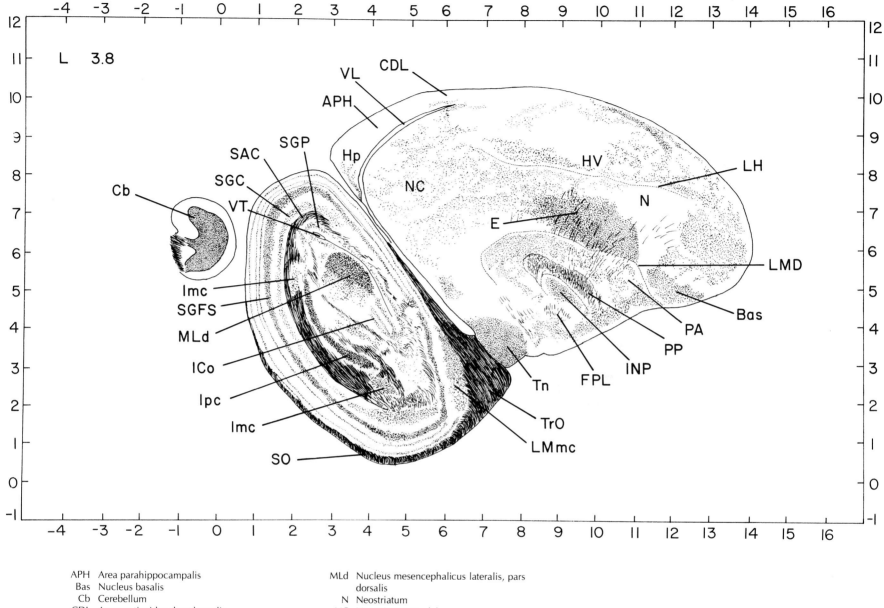

APH Area parahippocampalis
Bas Nucleus basalis
Cb Cerebellum
CDL Area corticoidea dorsolateralis
E Ectostriatum
FPL Fasciculus prosencephali lateralis (Lateral
forebrain bundle)
Hp Hippocampus
HV Hyperstriatum ventrale
ICo Nucleus intercollicularis
Imc Nucleus isthmi, pars magnocellularis
INP Nucleus intrapeduncularis
Ipc Nucleus isthmi, pars parvocellularis
LH Lamina hyperstriatica
LMD Lamina medullaris dorsalis
LMmc Nucleus lentiformis mesencephali, pars
magnocellularis

MLd Nucleus mesencephalicus lateralis, pars
dorsalis
N Neostriatum
NC Neostriatum caudale
PA Paleostriatum augmentatum (Caudate putamen)
PP Paleostriatum primitivum (Globus pallidus)
SAC Stratum album centrale
SGC Stratum griseum centrale
SGFS Stratum griseum et fibrosum superficiale
SGP Stratum griseum periventriculare
SO Stratum opticum
Tn Nucleus taeniae
TrO Tractus opticus
VL Ventriculus lateralis
VT Ventriculus tecti mesencephali

PLATE L 3.4   135

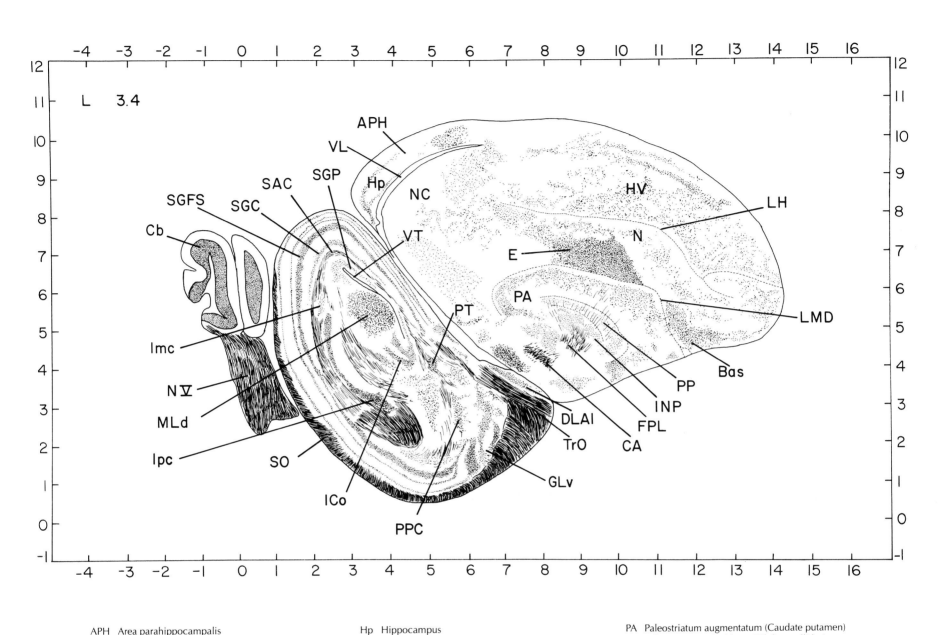

L 3.4

APH Area parahippocampalis
Bas Nucleus basalis
CA Commissura anterior [rostralis] (Anterior commissure)
Cb Cerebellum
DLAl Nucleus dorsolateralis anterior [rostralis] thalami, pars lateralis
E Ectostriatum
FPL Fasciculus prosencephali lateralis (Lateral forebrain bundle)
GLv Nucleus geniculatus lateralis, pars ventralis

Hp Hippocampus
HV Hyperstriatum ventrale
ICo Nucleus intercollicularis
Imc Nucleus isthmi, pars magnocellularis
INP Nucleus intrapeduncularis
Ipc Nucleus isthmi, pars parvocellularis
LH Lamina hyperstriatica
LMD Lamina medullaris dorsalis
MLd Nucleus mesencephalicus lateralis, pars dorsalis
N V Nervus trigeminus
N Neostriatum
NC Neostriatum caudale

PA Paleostriatum augmentatum (Caudate putamen)
PP Paleostriatum primitivum (Globus pallidus)
PPC Nucleus principalis precommissuralis
PT Nucleus pretectalis
SAC Stratum album centrale
SGC Stratum griseum centrale
SGFS Stratum griseum et fibrosum superficiale
SGP Stratum griseum periventriculare
SO Stratum opticum
TrO Tractus opticus
VL Ventriculus lateralis
VT Ventriculus tecti mesencephali

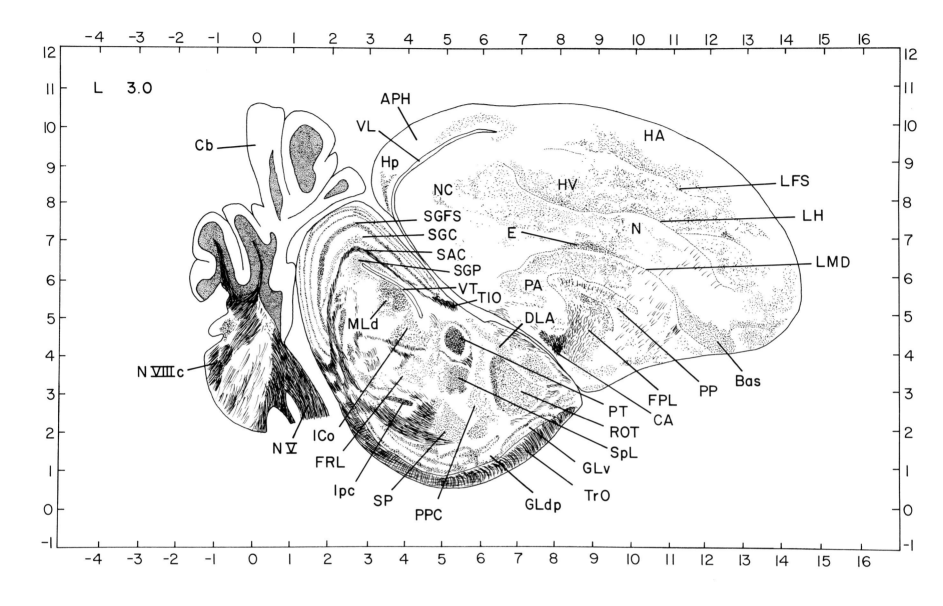

APH	Area parahippocampalis	Hp	Hippocampus	PP	Paleostriatum primitivum (Globus pallidus)
Bas	Nucleus basalis	HV	Hyperstriatum ventrale	PPC	Nucleus principalis precommissuralis
CA	Commissura anterior [rostralis] (Anterior commissure)	ICo	Nucleus intercollicularis	PT	Nucleus pretectalis
		Ipc	Nucleus isthmi, pars parvocellularis	ROT	Nucleus rotundus
Cb	Cerebellum	LFS	Lamina frontalis superior	SAC	Stratum album centrale
DLA	Nucleus dorsolateralis anterior [rostralis] thalami	LH	Lamina hyperstriatica	SGC	Stratum griseum centrale
		LMD	Lamina medullaris dorsalis	SGFS	Stratum griseum et fibrosum superficiale
E	Ectostriatum	MLd	Nucleus mesencephalicus lateralis, pars dorsalis	SGP	Stratum griseum periventriculare
FPL	Fasciculus prosencephali lateralis (Lateral forebrain bundle)			SP	Nucleus subpretectalis
		N	Neostriatum	SpL	Nucleus spiriformis lateralis
FRL	Formatio reticularis lateralis mesencephali	NC	Neostriatum caudale	TIO	Tractus isthmo-opticus
GLdp	Nucleus geniculatus lateralis, pars dorsalis principalis	N V	Nervus trigeminus	TrO	Tractus opticus
		N VIII c	Nervus octavus, pars cochlearis	VL	Ventriculus lateralis
GLv	Nucleus geniculatus lateralis, pars ventralis	PA	Paleostriatum augmentatum (Caudate putamen)	VT	Ventriculus tecti mesencephali
HA	Hyperstriatum accessorium				

PLATE L 2.6   137

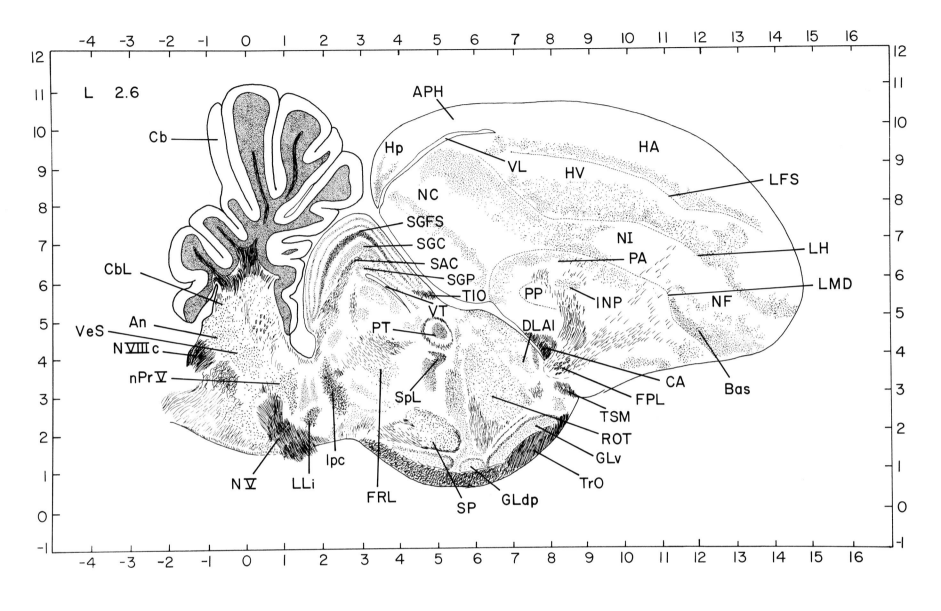

L 2.6

An	Nucleus angularis
APH	Area parahippocampalis
Bas	Nucleus basalis
CA	Commissura anterior [rostralis] (Anterior commissure)
Cb	Cerebellum
CbL	Nucleus cerebellaris lateralis
DLA	Nucleus dorsolateralis anterior [rostralis] thalami
FPL	Fasciculus prosencephali lateralis (Lateral forebrain bundle)
FRL	Formatio reticularis lateralis mesencephali
GLdp	Nucleus geniculatus lateralis, pars dorsalis principalis
GLv	Nucleus geniculatus lateralis, pars ventralis
HA	Hyperstriatum accessorium

Hp	Hippocampus
HV	Hyperstriatum ventrale
Ipc	Nucleus isthmi, pars parvocellularis
INP	Nucleus intrapeduncularis
LFS	Lamina frontalis superior
LH	Lamina hyperstriatica
LLi	Nucleus lemnisci lateralis, pars intermedia
LMD	Lamina medullaris dorsalis
N V	Nervus trigeminus
N VIII c	Nervus octavus, pars cochlearis
NC	Neostriatum caudale
NF	Neostriatum frontale
NI	Neostriatum intermedium
nPr V	Nucleus sensorius principalis nervi trigemini
PA	Paleostriatum augmentatum (Caudate putamen)

PP	Paleostriatum primitivum (Globus pallidus)
PT	Nucleus pretectalis
ROT	Nucleus rotundus
SAC	Stratum album centrale
SGC	Stratum griseum centrale
SGFS	Stratum griseum et fibrosum superficiale
SGP	Stratum griseum periventriculare
SP	Nucleus subpretectalis
SpL	Nucleus spiriformis lateralis
TIO	Tractus isthmo-opticus
TrO	Tractus opticus
TSM	Tractus septomesencephalicus
VeS	Nucleus vestibularis superior
VL	Ventriculus lateralis
VT	Ventriculus tecti mesencephali

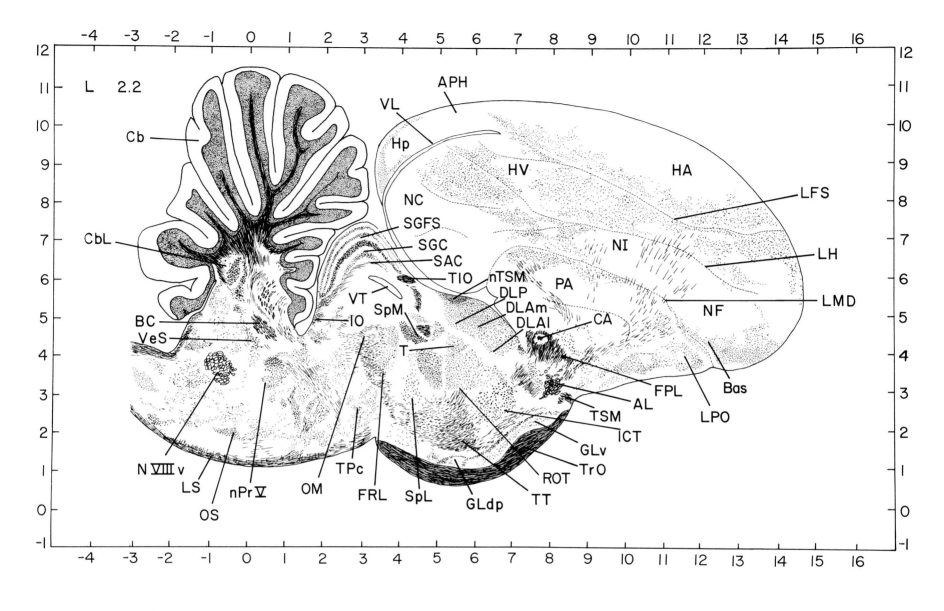

AL	Ansa lenticularis	
APH	Area parahippocampalis	
Bas	Nucleus basalis	
BC	Brachium conjunctivum	
Cb	Cerebellum	
CbL	Nucleus cerebellaris lateralis	
DLAl	Nucleus dorsolateralis anterior [rostralis] thalami, pars lateralis	
DLAm	Nucleus dorsolateralis anterior [rostralis] thalami, pars medialis	
DLP	Nucleus dorsolateralis posterior thalami	
FRL	Formatio reticularis lateralis mesencephali	
GLdp	Nucleus geniculatus lateralis, pars dorsalis principalis	
GLv	Nucleus geniculatus lateralis, pars ventralis	
HA	Hyperstriatum accessorium	
Hp	Hippocampus	
HV	Hyperstriatum ventrale	

ICT	Nucleus intercalatus thalami	
IO	Nucleus isthmo-opticus	
FPL	Fasciculus prosencephali lateralis (Lateral forebrain bundle)	
LFS	Lamina frontalis superior	
LH	Lamina hyperstriatica	
LMD	Lamina medullaris dorsalis	
LPO	Lobus parolfactorius	
LS	Lemniscus spinalis	
N VIII v	Nervus octavus, pars vestibularis	
NC	Neostriatum caudale	
NF	Neostriatum frontale	
NI	Neostriatum intermedium	
nPr V	Nucleus sensorius principalis nervi trigemini	
nTSM	Nucleus tractus septomesencephalicus (Nucleus superficialis parvocellularis)	
OM	Tractus occipitomesencephalicus	
OS	Nucleus olivaris superior	

PA	Paleostriatum augmentatum (Caudate putamen)	
ROT	Nucleus rotundus	
SAC	Stratum album centrale	
SGC	Stratum griseum centrale	
SGFS	Stratum griseum et fibrosum superficiale	
SpL	Nucleus spiriformis lateralis	
SpM	Nucleus spiriformis medialis	
T	Nucleus triangularis	
TIO	Tractus isthmo-opticus	
TPc	Nucleus tegmenti pedunculo-pontinus, pars compacta (Substantia nigra)	
TrO	Tractus opticus	
TSM	Tractus septomesencephalicus	
TT	Tractus tectothalamicus	
VeS	Nucleus vestibularis superior	
VL	Ventriculus lateralis	
VT	Ventriculus tecti mesencephali	

PLATE L 1.8

139

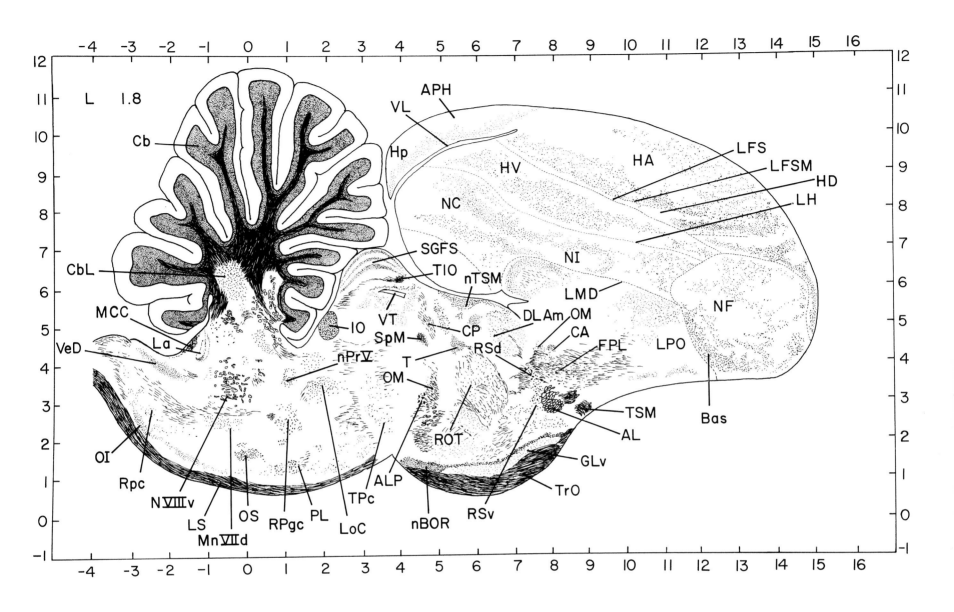

AL   Ansa lenticularis
ALP   Nucleus ansae lenticularis posterior
      [caudalis]
APH   Area parahippocampalis
Bas   Nucleus basalis
CA   Commissura anterior [rostralis] (Anterior
      commissure)
Cb   Cerebellum
CbL   Nucleus cerebellaris lateralis
CP   Commissura posterior [caudalis] (Posterior
      commissure)
DLAm Nucleus dorsolateralis anterior [rostralis]
      thalami, pars medialis
FPL   Fasciculus prosencephali lateralis (Lateral
      forebrain bundle)
GLv   Nucleus geniculatus lateralis, pars ventralis
HA   Hyperstriatum accessorium
HD   Hyperstriatum dorsale
Hp   Hippocampus
HV   Hyperstriatum ventrale
IO   Nucleus isthmo-opticus
La   Nucleus laminaris

LFS   Lamina frontalis superior
LFSM Lamina frontalis suprema
LH   Lamina hyperstriatica
LMD   Lamina medullaris dorsalis
LoC   Locus ceruleus
LPO   Lobus parolfactorius
LS   Lemniscus spinalis
MCC   Nucleus magnocellularis cochlearis
Mn VII d Nucleus motorius nervi facialis, pars dorsalis
N VIII v Nervus octavus, pars vestibularis
nBOR Nucleus opticus basalis; nucleus
      ectomamillaris (Nucleus of the basal optic
      root)
NC   Neostriatum caudale
NF   Neostriatum frontale
NI   Neostriatum intermedium
nPr V Nucleus sensorius principalis nervi trigemini
nTSM Nucleus tractus septomesencephalicus;
      nucleus superficialis parvocellularis
OI   Nucleus olivaris inferior

OM   Tractus occipitomesencephalicus
OS   Nucleus olivaris superior
PL   Nucleus pontis lateralis
ROT   Nucleus rotundus
Rpc   Nucleus reticularis parvocellularis
RPgc Nucleus reticularis pontis caudalis, pars
      gigantocellularis
RSd   Nucleus reticularis superior, pars dorsalis
RSv   Nucleus reticularis superior, pars ventralis
SGFS Stratum griseum et fibrosum superficiale
SpM   Nucleus spiriformis medialis
T   Nucleus triangularis
TIO   Tractus isthmo-opticus
TPc   Nucleus tegmenti pedunculo-pontinus, pars
      compacta (Substantia nigra)
TrO   Tractus opticus
TSM   Tractus septomesencephalicus
VeD   Nucleus vestibularis descendens
VL   Ventriculus lateralis
VT   Ventriculus tecti mesencephali

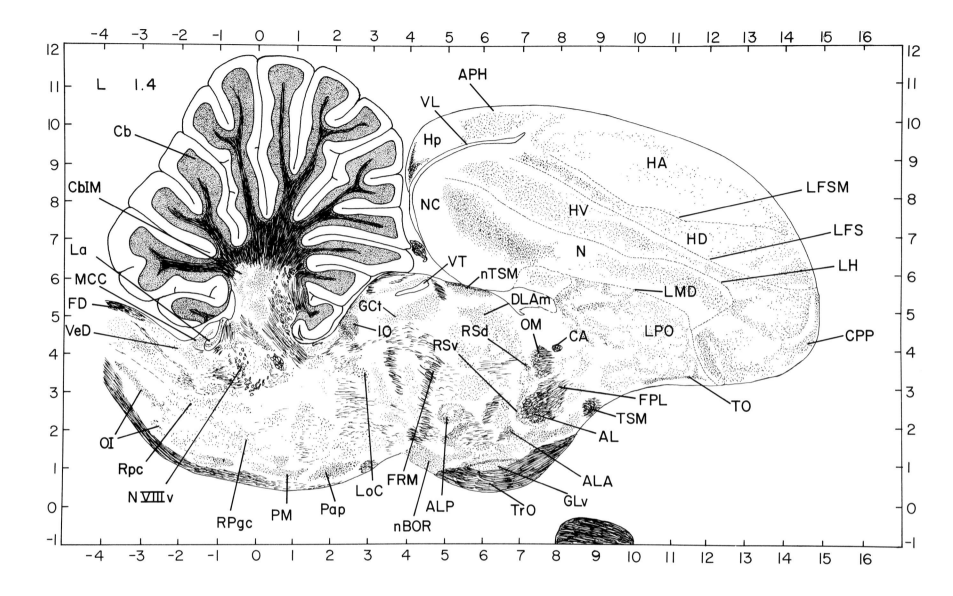

AL   Ansa lenticularis
ALA   Nucleus ansae lenticularis anterior [rostralis]
ALP   Nucleus ansae lenticularis posterior
      [caudalis]
APH   Area parahippocampalis
CA   Commissura anterior [rostralis] (Anterior
      commissure)
Cb   Cerebellum
CbIM Nucleus cerebellaris intermedius
CPP   Cortex prepiriformis
DLAm Nucleus dorsolateralis anterior [rostralis]
      thalami, pars medialis
FD   Funiculus dorsalis
FPL   Fasciculus prosencephali lateralis (Lateral
      forebrain bundle)
FRM   Formatio reticularis medialis mesencephali
GCT   Substantia grisea centralis
GLv   Nucleus geniculatus lateralis, pars ventralis

HA   Hyperstriatum accessorium
HD   Hyperstriatum dorsale
Hp   Hippocampus
HV   Hyperstriatum ventrale
IO   Nucleus isthmo-opticus
La   Nucleus laminaris
LFS   Lamina frontalis superior
LFSM   Lamina frontalis suprema
LH   Lamina hyperstriatica
LMD   Lamina medullaris dorsalis
LoC   Locus ceruleus
LPO   Lobus parolfactorius
MCC   Nucleus magnocellularis cochlearis
nBOR   Nucleus opticus basalis; nucleus
       ectomamillaris (Nucleus of the basal optic
       root)
NC   Neostriatum caudale
NI   Neostriatum intermedium

nTSM   Nucleus tractus septomesencephalicus;
       nucleus superficialis parvocellularis
N VIII v Nervus octavus, pars vestibularis
OI   Nucleus olivaris inferior
OM   Tractus occipitomesencephalicus
Pap   Nucleus papillioformis
PM   Nucleus pontis medialis
Rpc   Nucleus reticularis parvocellularis
RPgc   Nucleus reticularis pontis caudalis, pars
       gigantocellularis
RSd   Nucleus reticularis superior, pars dorsalis
RSv   Nucleus reticularis superior, pars ventralis
TO   Tuberculum olfactorium
TrO   Tractus opticus
TSM   Tractus septomesencephalicus
VeD   Nucleus vestibularis descendens
VL   Ventriculus lateralis
VT   Ventriculus tecti mesencephali

PLATE L 1.0   141

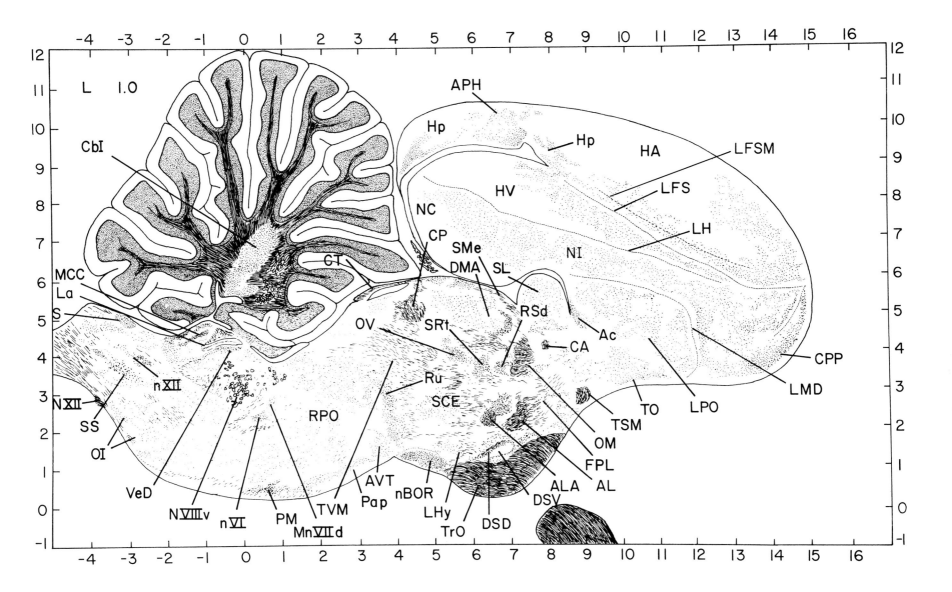

Ac	Nucleus accumbens	La	Nucleus laminaris	OI	Nucleus olivaris inferior

Ac Nucleus accumbens
AL Ansa lenticularis
ALA Nucleus ansae lenticularis anterior [rostralis]
APH Area parahippocampalis
AVT Area ventralis (Tsai)
CA Commissura anterior [rostralis] (Anterior commissure)
CbI Nucleus cerebellaris internus
CP Commissura posterior [caudalis] (Posterior commissure)
CPP Cortex prepiriformis
CT Commissura tectalis
DMA Nucleus dorsomedialis anterior thalami
DSD Decussatio supraoptica dorsalis
DSV Decussatio supraoptica ventralis
FPL Fasciculus prosencephali lateralis (Lateral forebrain bundle)
HA Hyperstriatum accessorium
Hp Hippocampus
HV Hyperstriatum ventrale

La Nucleus laminaris
LFS Lamina frontalis superior
LFSM Lamina frontalis suprema
LH Lamina hyperstriatica
LHy Regio lateralis hypothalami (Lateral hypothalamic area)
LMD Lamina medullaris dorsalis
LPO Lobus parolfactorius
MCC Nucleus magnocellularis cochlearis
Mn VII d Nucleus motorius nervi facialis, pars dorsalis
nBOR Nucleus opticus basalis; nucleus ectomamillaris (Nucleus of the basal optic root)
NC Neostriatum caudale
NI Neostriatum intermedium
n VI Nucleus nervi abducentis
N XII Nervus hypoglossus
N VIII v Nervus octavus, pars vestibularis
n XII Nucleus nervi hypoglossi
OM Tractus occipitomesencephalicus

OI Nucleus olivaris inferior
OV Nucleus ovoidalis
Pap Nucleus papillioformis
PM Nucleus pontis medialis
RPO Nucleus reticularis pontis oralis
RSd Nucleus reticularis superior, pars dorsalis
Ru Nucleus ruber
S Nucleus tractus solitarii
SCE Stratum cellulare externum
SL Nucleus septalis lateralis
SMe Stria medullaris
SRt Nucleus subrotundus
SS Nucleus supraspinalis
TO Tuberculum olfactorium
TrO Tractus opticus
TSM Tractus septomesencephalicus
TVM Tractus vestibulomesencephalicus (Papez)
VeD Nucleus vestibularis descendens
VL Ventriculus lateralis

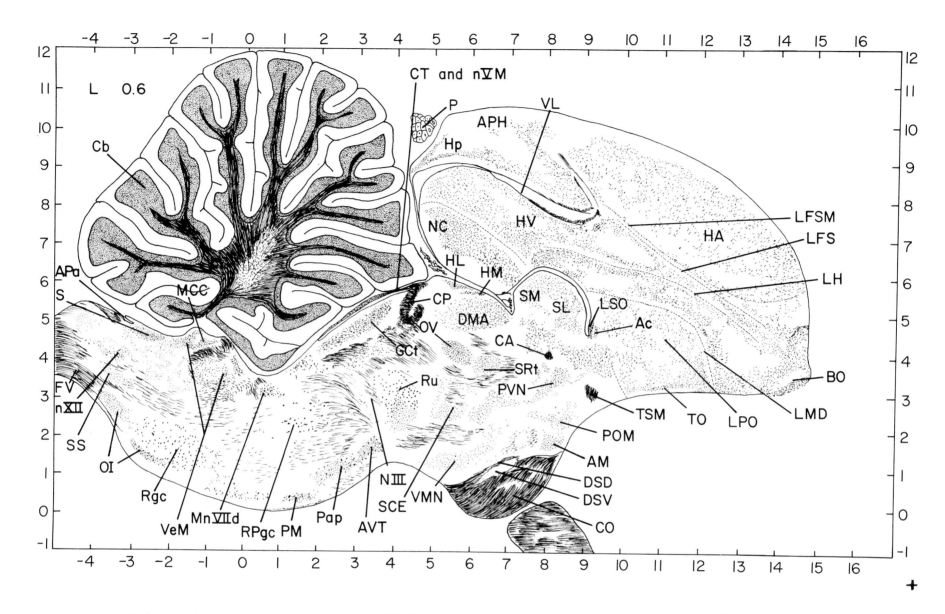

Ac	Nucleus accumbens	HA	Hyperstriatum accessorium
AM	Nucleus anterior [rostralis] medialis	HL	Nucleus habenularis lateralis
	hypothalami	HM	Nucleus habenularis medialis
APa	Area postrema	Hp	Hippocampus
APH	Area parahippocampalis	HV	Hyperstriatum ventrale
AVT	Area ventralis (Tsai)	LFS	Lamina frontalis superior
BO	Bulbus olfactorius	LFSM	Lamina frontalis suprema
CA	Commissura anterior [rostralis] (Anterior	LH	Lamina hyperstriatica
	commissure)	LMD	Lamina medullaris dorsalis
Cb	Cerebellum	LPO	Lobus parolfactorius
CO	Chiasma opticum	LSO	Organum septi laterale (Lateral septal organ)
CP	Commissura posterior [caudalis] (Posterior	MCC	Nucleus magnocellularis cochlearis
	commissure)	Mn VII d	Nucleus motorius nervi facialis, pars dorsalis
CT	Commissura tectalis	NC	Neostriatum caudale
DMA	Nucleus dorsomedialis anterior [rostralis]	N III	Nervus oculomotorius
	thalami	n V M	Nucleus mesencephalicus nervi trigemini
DSD	Decussatio supraoptica dorsalis	n XII	Nucleus nervi hypoglossi
DSV	Decussatio supraoptica ventralis	OI	Nucleus olivaris inferior
FV	Funiculus ventralis	OV	Nucleus ovoidalis
GCt	Substantia grisea centralis	P	Glandula pinealis (Pineal gland)

Pap	Nucleus papillioformis
PM	Nucleus pontis medialis
POM	Nucleus preopticus medialis
PVN	Nucleus paraventricularis magnocellularis
	(Paraventricular nucleus)
Rgc	Nucleus reticularis gigantocellularis
RPgc	Nucleus reticularis pontis caudalis, pars
	gigantocellularis
Ru	Nucleus ruber
S	Nucleus tractus solitarii
SCE	Stratum cellulare externum
SL	Nucleus septalis lateralis
SM	Nucleus septalis medialis
SRt	Nucleus subrotundus
TO	Tuberculum olfactorium
TSM	Tractus septomesencephalicus
VeM	Nucleus vestibularis medialis
VL	Ventriculus lateralis
VMN	Nucleus ventromedialis hypothalami

PLATE L 0.2   143

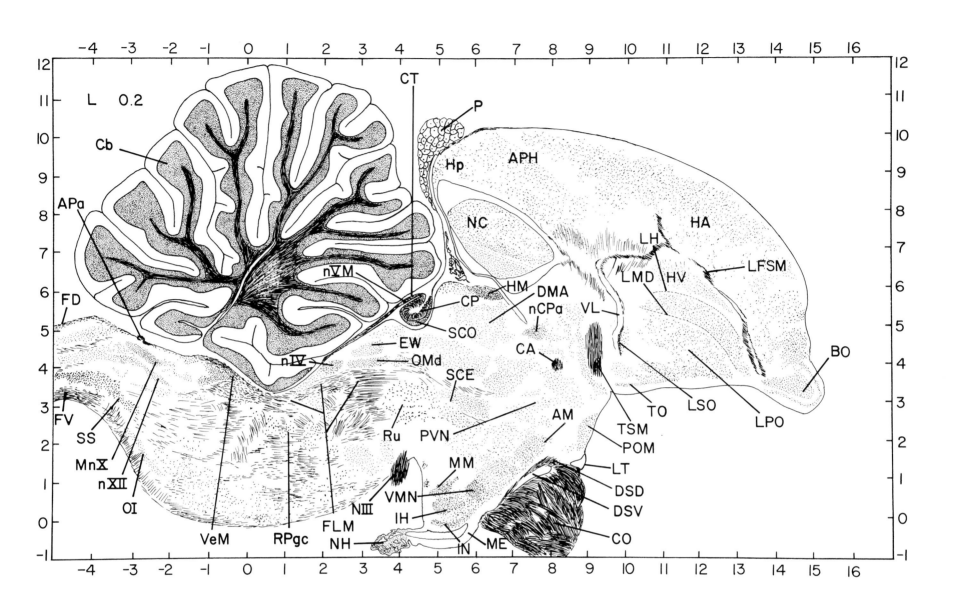

AM	Nucleus anterior [rostralis] medialis hypothalami	
APa	Area postrema	
APH	Area parahippocampalis	
BO	Bulbus olfactorius	
CA	Commissura anterior [rostralis] (Anterior commissure)	
Cb	Cerebellum	
CO	Chiasma opticum	
CP	Commissura posterior [caudalis] (Posterior commissure)	
CT	Commissura tectalis	
DMA	Nucleus dorsomedialis anterior [rostralis] thalami	
DSD	Decussatio supraoptica dorsalis	
DSV	Decussatio supraoptica ventralis	
EW	Nucleus of Edinger-Westphal	
FD	Funiculus dorsalis	
FLM	Fasciculus longitudinalis medialis	
FV	Funiculus ventralis	

HA	Hyperstriatum accessorium
HM	Nucleus habenularis medialis
Hp	Hippocampus
HV	Hyperstriatum ventrale
IH	Nucleus inferioris hypothalami
IN	Nucleus infundibuli hypothalami
LFSM	Lamina frontalis suprema
LH	Lamina hyperstriatica
LMD	Lamina medullaris dorsalis
LPO	Lobus parolfactorius
LSO	Organum septi laterale (Lateral septal organ)
LT	Lamina terminalis
ME	Eminentia mediana (Median eminence)
MM	Nucleus mamillaris medialis
Mn X	Nucleus motorius dorsalis nervi vagi
n V M	Nucleus mesencephalicus nervi trigemini
NC	Neostriatum caudale
nCPa	Nucleus commissurae pallii (Bed nucleus pallial commissure)
NH	Neurohypophysis

N III	Nervus oculomotorius
n IV	Nucleus nervi trochlearis
n XII	Nucleus nervi hypoglossi
OI	Nucleus olivaris inferior
OMd	Nucleus nervi oculomotorii, pars dorsalis
P	Glandula pinealis (Pineal gland)
POM	Nucleus preopticus medialis
PVN	Nucleus paraventricularis magnocellularis (Paraventricular nucleus)
RPgc	Nucleus reticularis pontis caudalis, pars gigantocellularis
Ru	Nucleus ruber
SCE	Stratum cellulare externum
SCO	Organum subcommissurale (Subcommissural organ)
TO	Tuberculum olfactorium
TSM	Tractus septomesencephalicus
VL	Ventriculus lateralis
VeM	Nucleus vestibularis medialis
VMN	Nucleus ventromedialis hypothalami

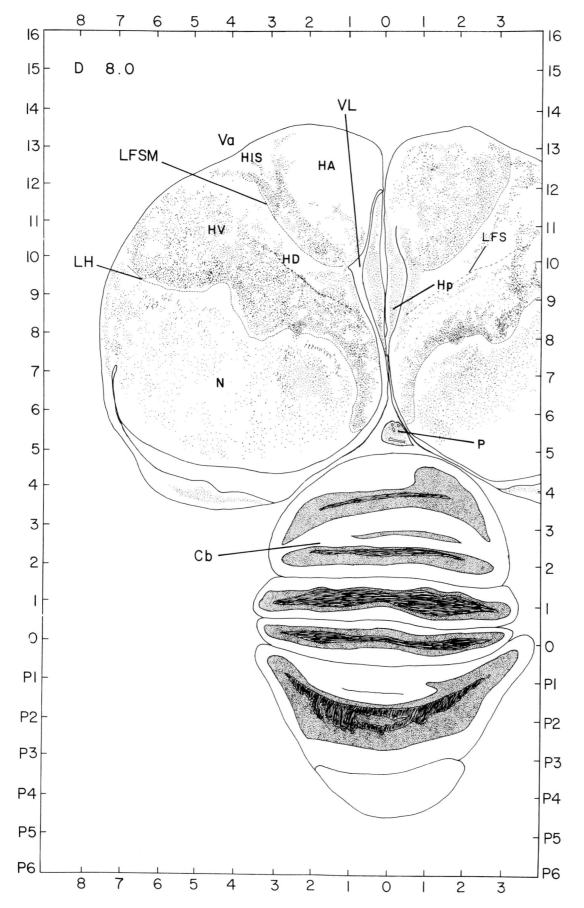

Cb   Cerebellum
HA   Hyperstriatum accessorium
HD   Hyperstriatum dorsale
HIS   Hyperstriatum intercalatum supremum
Hp   Hippocampus
HV   Hyperstriatum ventrale
LFS   Lamina frontalis superior
LFSM   Lamina frontalis suprema
LH   Lamina hyperstriatica
N   Neostriatum
P   Glandula pinealis (Pineal gland)
Va   Vallecula telencephali
VL   Ventriculus lateralis

Cb   Cerebellum
HA   Hyperstriatum accessorium
HD   Hyperstriatum dorsale
HIS Hyperstriatum intercalatum supremum
Hp   Hippocampus
HV   Hyperstriatum ventrale
LFS Lamina frontalis superior
LFSM Lamina frontalis suprema
LH   Lamina hyperstriatica
N   Neostriatum
P   Glandula pinealis (Pineal gland)
SGFS Stratum griseum et fibrosum superficiale
Va   Vallecula telencephali
VL   Ventriculus lateralis

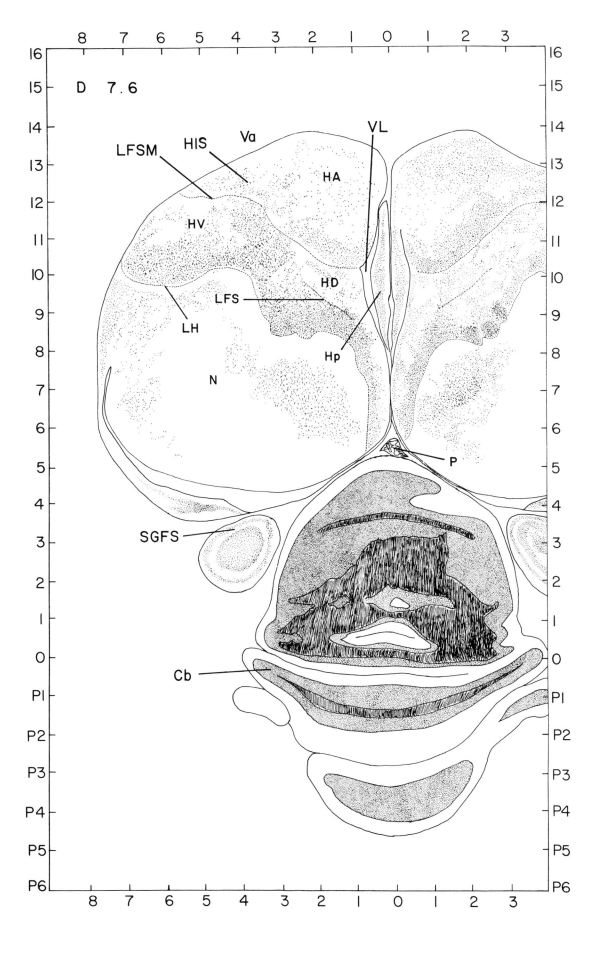

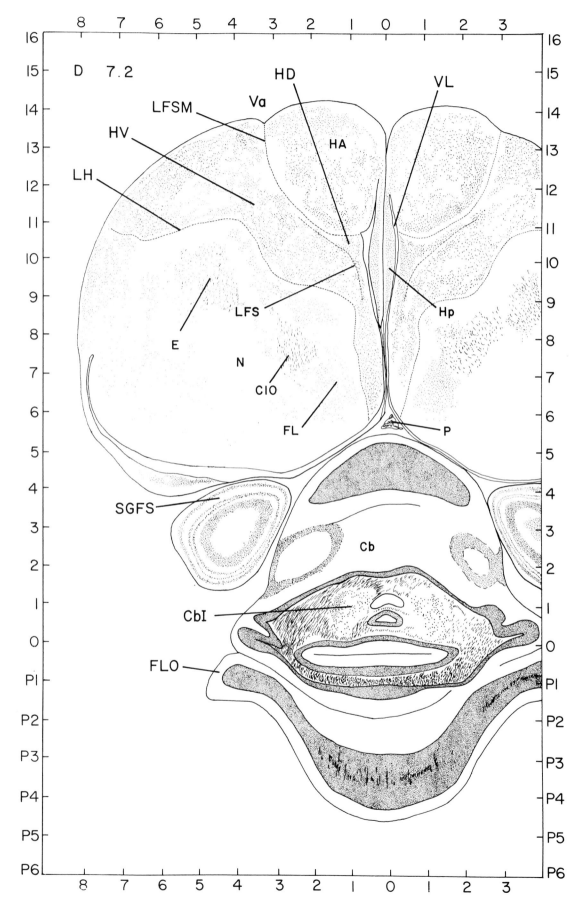

Cb   Cerebellum
Cbl   Nucleus cerebellaris internus
CIO   Capsula interna occipitalis
E   Ectostriatum
FL   Field L
FLO   Flocculus
HA   Hyperstriatum accessorium
HD   Hyperstriatum dorsale
HV   Hyperstriatum ventrale
Hp   Hippocampus
LFS   Lamina frontalis superior
LFSM   Lamina frontalis suprema
LH   Lamina hyperstriatica
N   Neostriatum
P   Glandula pinealis (Pineal gland)
SGFS   Stratum griseum et fibrosum superficiale
Va   Vallecula telencephali
VL   Ventriculus lateralis

CbI Nucleus cerebellaris internus
CIO Capsula interna occipitalis
E Ectostriatum
FL Field L
FLO Flocculus
HA Hyperstriatum accessorium
HD Hyperstriatum dorsale
HV Hyperstriatum ventrale
LFS Lamina frontalis superior
LFSM Lamina frontalis suprema
LH Lamina hyperstriatica
N Neostriatum
P Glandula pinealis (Pineal gland)
SAC Stratum album centrale
SGFS Stratum griseum et fibrosum superficiale
Va Vallecula telencephali
VC Ventriculus cerebelli
VL Ventriculus lateralis

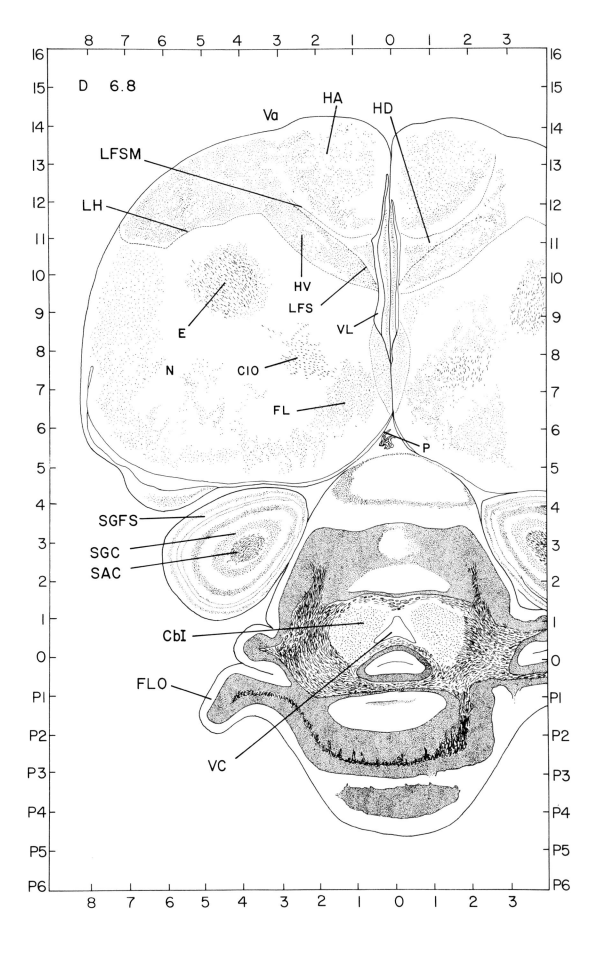

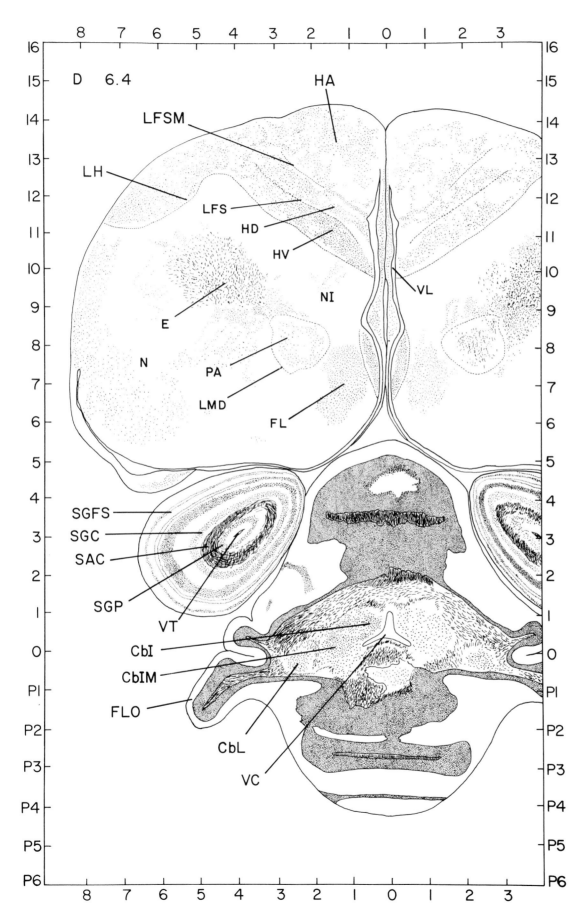

D 6.4

CbI   Nucleus cerebellaris internus
CbL   Nucleus cerebellaris lateralis
CbIM   Nucleus cerebellaris intermedius
E   Ectostriatum
FL   Field L
FLO   Flocculus
HA   Hyperstriatum accessorium
HD   Hyperstriatum dorsale
HV   Hyperstriatum ventrale
LFS   Lamina frontalis superior
LFSM   Lamina frontalis suprema
LH   Lamina hyperstriatica
LMD   Lamina medullaris dorsalis
N   Neostriatum
NI   Neostriatum intermedium
PA   Paleostriatum augmentatum (Caudate putamen)
SAC   Stratum album centrale
SGC   Stratum griseum centrale
SGFS   Stratum griseum et fibrosum superficiale
SGP   Stratum griseum periventriculare
VC   Ventriculus cerebelli
VL   Ventriculus lateralis
VT   Ventriculus tecti mesencephali

CbI   Nucleus cerebellaris internus
CbL   Nucleus cerebellaris lateralis
CbIM   Nucleus cerebellaris intermedius
CbIvm Nucleus cerebellaris internus, pars
       ventromedialis
E   Ectostriatum
FLO   Flocculus
HA   Hyperstriatum accessorium
HD   Hyperstriatum dorsale
HIS   Hyperstriatum intercalatum supremum
HV   Hyperstriatum ventrale
LFS   Lamina frontalis superior
LFSM   Lamina frontalis suprema
LH   Lamina hyperstriatica
LMD   Lamina medullaris dorsalis
N   Neostriatum
NI   Neostriatum intermedium
PA   Paleostriatum augmentatum (Caudate
       putamen)
PCV III Plexus choroideus ventriculi tertii (Choroid
       plexus within third ventricle)
PP   Paleostriatum primitivum (Globus pallidus)
SAC   Stratum album centrale
SGC   Stratum griseum centrale
SGFS   Stratum griseum et fibrosum superficiale
SGP   Stratum griseum periventriculare
Va   Vallecula telencephali
VC   Ventriculus cerebelli
VL   Ventriculus lateralis
VT   Ventriculus tecti mesencephali

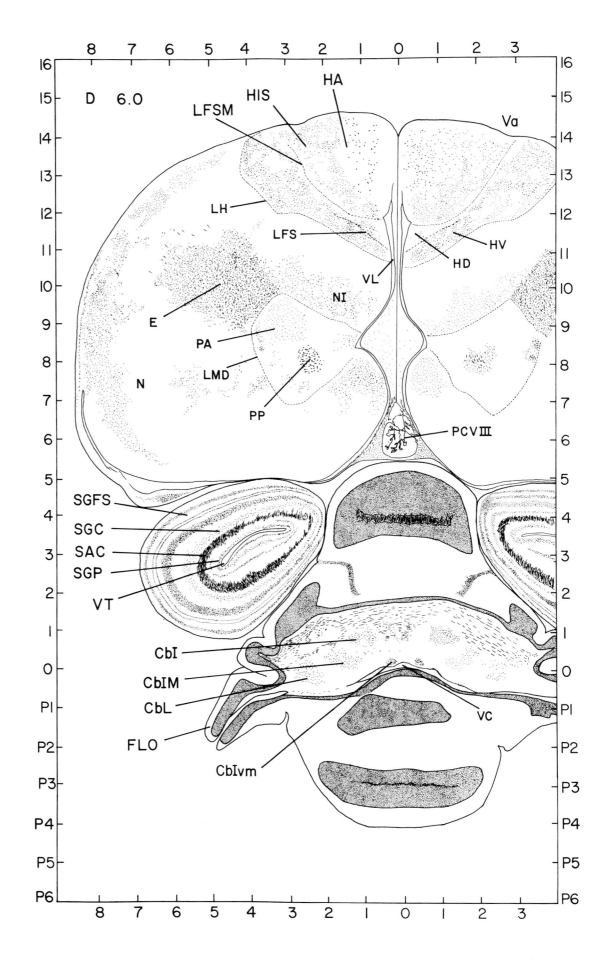

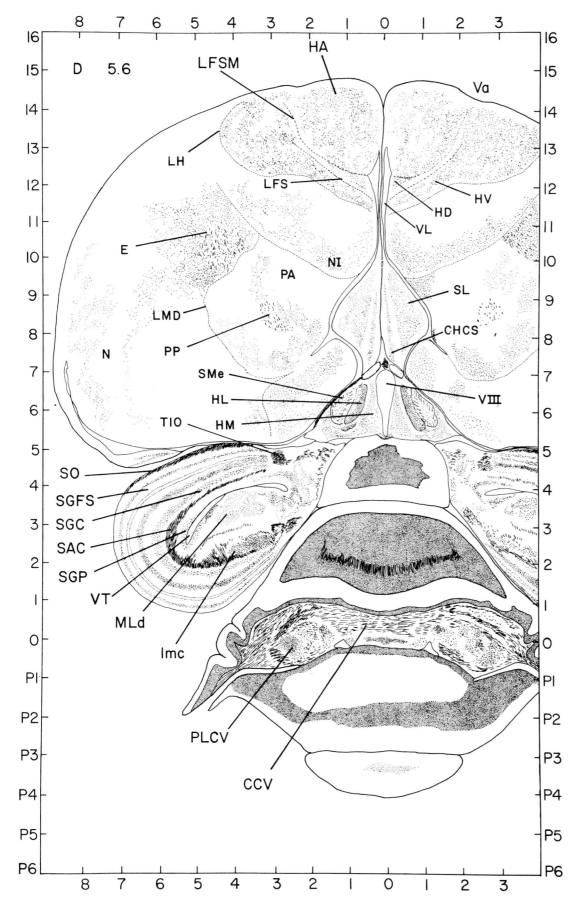

CCV   Commissura cerebellaris ventralis
CHCS Tractus cortico-habenularis et cortico-septalis
E   Ectostriatum
HA   Hyperstriatum accessorium
HD   Hyperstriatum dorsale
HL   Nucleus habenularis lateralis
HM   Nucleus habenularis medialis
HV   Hyperstriatum ventrale
Imc   Nucleus isthmi, pars magnocellularis
LFS   Lamina frontalis superior
LFSM Lamina frontalis suprema
LH   Lamina hyperstriatica
LMD   Lamina medullaris dorsalis
MLd   Nucleus mesencephalicus lateralis, pars
      dorsalis
N   Neostriatum
NI   Neostriatum intermedium
PA   Paleostriatum augmentatum (Caudate putamen)
PLCV Processus lateralis cerebello-vestibularis
PP   Paleostriatum primitivum (Globus pallidus)
SAC   Stratum album centrale
SGC   Stratum griseum centrale
SGFS Stratum griseum et fibrosum superficiale
SGP   Stratum griseum periventriculare
SL   Nucleus septalis lateralis
SMe   Stria medullaris
SO   Stratum opticum
TIO   Tractus isthmo-opticus
Va   Vallecula telencephali
VL   Ventriculus lateralis
VT   Ventriculus tecti mesencephali
V III Ventriculus tertius (Third ventricle)

AId Archistriatum intermedium, pars dorsalis (Zeier and Karten)
BC Brachium conjunctivum
CP Commissura posterior [caudalis] (Posterior commissure)
CPi Cortex piriformis
CHCS Tractus cortico-habenularis et cortico-septalis
CT Commissura tectalis
DMA Nucleus dorsomedialis anterior [rostralis] thalami
DLA Nucleus dorsolateralis anterior [rostralis] thalami
E Ectostriatum
HA Hyperstriatum accessorium
HD Hyperstriatum dorsale
HV Hyperstriatum ventrale
ICo Nucleus intercollicularis
Imc Nucleus isthmi, pars magnocellularis
Ipc Nucleus isthmi, pars parvocellularis
LAD Lamina archistriatalis dorsalis
LFS Lamina frontalis superior
LFSM Lamina frontalis suprema
LH Lamina hyperstriatica
LMD Lamina medullaris dorsalis
LSO Organum septi laterale (Lateral septal organ)
MLd Nucleus mesencephalicus lateralis, pars dorsalis
N Neostriatum
n V m Nucleus mesencephalicus nervi trigemini
PA Paleostriatum augmentatum (Caudate putamen)
PP Paleostriatum primitivum (Globus pallidus)
SAC Stratum album centrale
SCO Organum subcommissurale (Subcommissural organ)
SGC Stratum griseum centrale
SGFS Stratum griseum et fibrosum superficiale
SGP Stratum griseum periventriculare
SL Nucleus septalis lateralis
SMe Stria medullaris
SO Stratum opticum
TIO Tractus isthmo-opticus
TSM Tractus septomesencephalicus
VL Ventriculus lateralis
VT Ventriculus tecti mesencephali
V III Ventriculus tertius (Third ventricle)

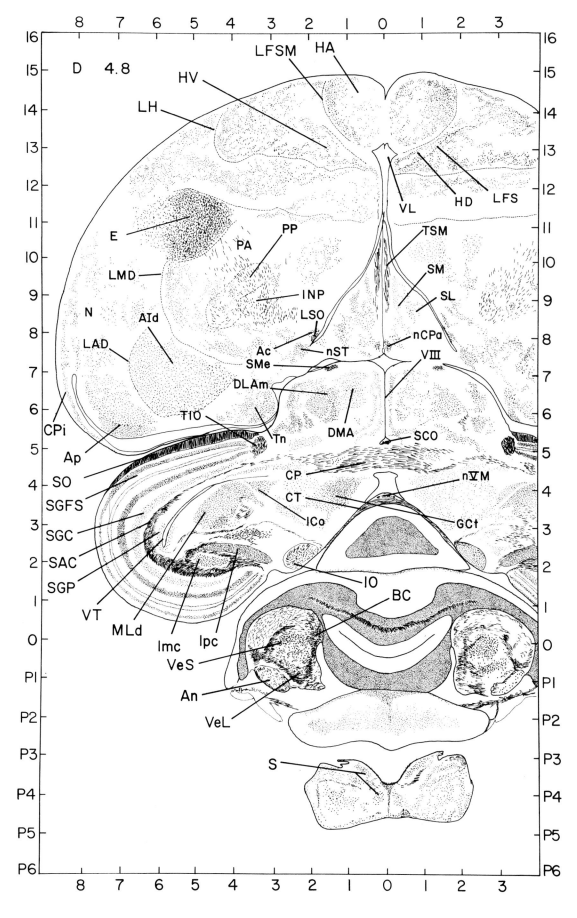

D 4.8

AA   Archistriatum anterior [rostrale] (Zeier and
       Karten)
Ac   Nucleus accumbens
Am   Archistriatum mediale (Zeier and Karten)
An   Nucleus angularis
AIv Archistriatum intermedium, pars ventralis
       (Zeier and Karten)
Ap   Archistriatum posterior [caudale] (Zeier and
       Karten)
Bas Nucleus basalis
CPi Cortex piriformis
DLAm Nucleus dorsolateralis anterior [rostralis]
       thalami, pars medialis
DLP Nucleus dorsolateralis posterior [caudalis]
       thalami
DMA Nucleus dorsomedialis anterior [rostralis]
       thalami
DMP Nucleus dorsomedialis posterior [caudalis]
       thalami
D IV Decussatio nervi trochlearis
E   Ectostriatum
EW   Nucleus of Edinger-Westphal
FA   Tractus fronto-archistriaticus
FPL Fasciculus prosencephali lateralis (Lateral
       forebrain bundle)
GCt Substantia grisea centralis
HA   Hyperstriatum accessorium
HV   Hyperstriatum ventrale
ICo Nucleus intercollicularis
Imc Nucleus isthmi, pars magnocellularis
INP Nucleus intrapeduncularis
IO   Nucleus isthmo-opticus
Ipc Nucleus isthmi, pars parvocellularis
La   Nucleus laminaris
LAD Lamina archistriatalis dorsalis
LH   Lamina hyperstriatica
LMD Lamina medullaris dorsalis
LSO Organum septi laterale (Lateral septal organ)
MCC Nucleus magnocellularis cochlearis
MLd Nucleus mesencephalicus lateralis, pars
       dorsalis
Mn X Nucleus motorius dorsalis nervi vagi
nCPa Nucleus commissurae pallii (Bed nucleus
       pallial commissure)
nTSM Nucleus tractus septomesencephalicus
       (Nucleus superficialis parvocellularis)
n XII Nucleus nervi hypoglossi
PA   Paleostriatum augmentatum (Caudate putamen)
PP   Paleostriatum primitivum (Globus pallidus)
S   Nucleus tractus solitarii
SAC Stratum album centrale
SGC Stratum griseum centrale
SGFS Stratum griseum et fibrosum superficiale
SGP Stratum griseum periventriculare
SL   Nucleus septalis lateralis
SM   Nucleus septalis medialis
SMe Stria medullaris
SO   Stratum opticum
SSO Organum subseptale (Subseptal, subfornical or
       interventricular organ)
TIO Tractus isthmo-opticus
Tn   Nucleus taeniae
TSM Tractus septomesencephalicus
TVM Tractus vestibulomesencephalicus
VeL Nucleus vestibularis lateralis
VeM Nucleus vestibularis medialis
VeS Nucleus vestibularis superior
VL   Ventriculus lateralis
VT   Ventriculus tecti mesencephali
V III Ventriculus tertius (Third ventricle)

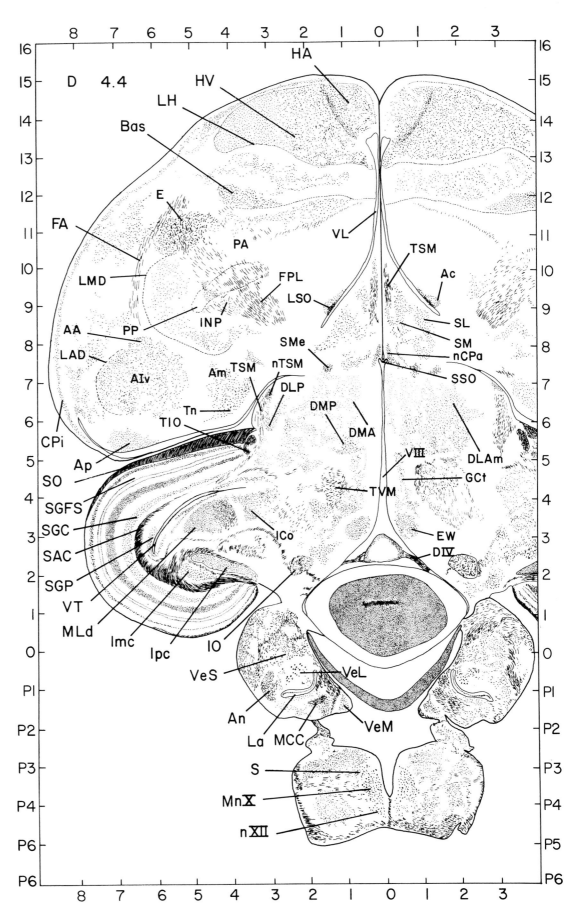

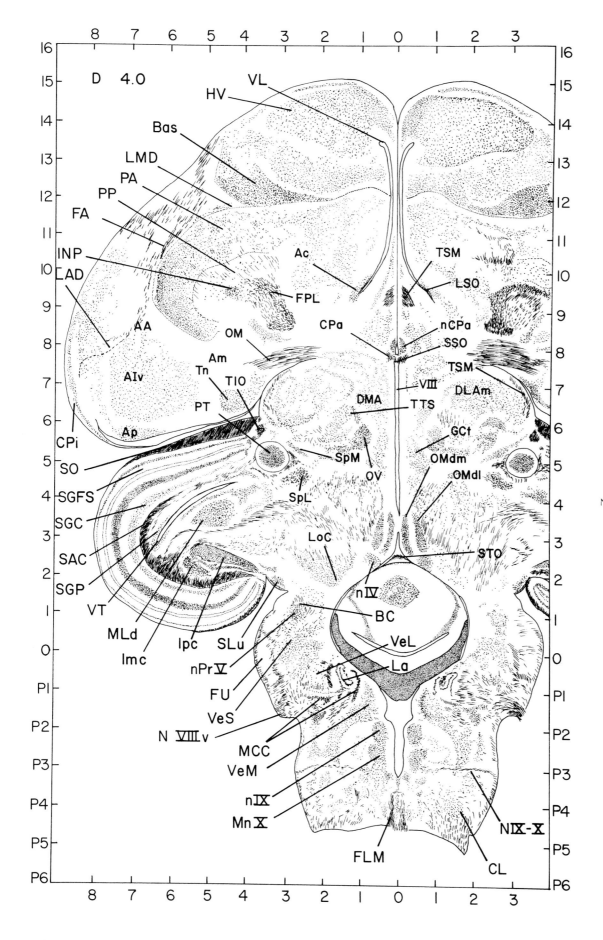

D 4.0

AA Architriatum anterior [rostrale] (Zeier and
    Karten)
Ac Nucleus accumbens
AIv Archistriatum intermedium, pars ventralis
    (Zeier and Karten)
Am Archistriatum mediale (Zeier and Karten)
Bas Nucleus basalis
BC Brachium conjunctivum
CA Commissura anterior [rostralis] (Anterior
    commissure)
CPi Cortex piriformis
CPP Cortex prepiriformis
D   Nucleus of Darkschewitsch
DLAl Nucleus dorsolateralis anterior [rostralis]
     thalami, pars lateralis
DLAm Nucleus dorsolateralis anterior [rostralis]
     thalami, pars medialis
FDB Fasciculus diagonalis Brocae
FLM Fasciculus longitudinalis medialis
FPL Fasciculus prosencephali lateralis (Lateral
    forebrain bundle)
FU Fasciculus uncinatus (Russell)
Imc Nucleus isthmi, pars magnocellularis
Ipc Nucleus isthmi, pars parvocellularis
LAD Lamina archistriatalis dorsalis
LMD Lamina medullaris dorsalis
LoC Locus ceruleus
LPO Lobus parolfactorius
LSO Organum septi laterale (Lateral septal organ)
Mn X Nucleus motorius dorsalis nervi vagi
nCPa Nucleus commissurae pallii (Bed nucleus
     pallial commissure)
nPr V Nucleus sensorius principalis nervi trigemini
n IV Nucleus nervi trochlearis
n IX Nucleus nervi glossopharyngei
n XII Nucleus nervi hypoglossi
N VIII v Nervus octavus, pars vestibularis
OM Tractus occipitomesencephalicus
OMv Nucleus nervi oculomotorii, pars ventralis
OV Nucleus ovoidalis
PA Paleostriatum augmentatum (Caudate
    putamen)
PMI Nucleus paramedianus internus thalami
PT Nucleus pretectalis
RST Nucleus reticularis subtrigeminalis
SAC Stratum album centrale
SGC Stratum griseum centrale
SGFS Stratum griseum et fibrosum superficiale
SGP Stratum griseum periventriculare
SLu Nucleus semilunaris
SO Stratum opticum
SpL Nucleus spiriformis lateralis
SS Nucleus supraspinalis
SSO Organum subseptale (Subseptal, subfornical
    or interventricular organ)
T   Nucleus triangularis
Tn Nucleus taeniae
TIO Tractus isthmo-opticus
TSM Tractus septomesencephalicus
TTS Tractus thalamostriaticus
VeD Nucleus vestibularis descendens
VeL Nucleus vestibularis lateralis
VeM Nucleus vestibularis medialis
VL Ventriculus lateralis
VT Ventriculus tecti mesencephali
V III Ventriculus tertius (Third ventricle)

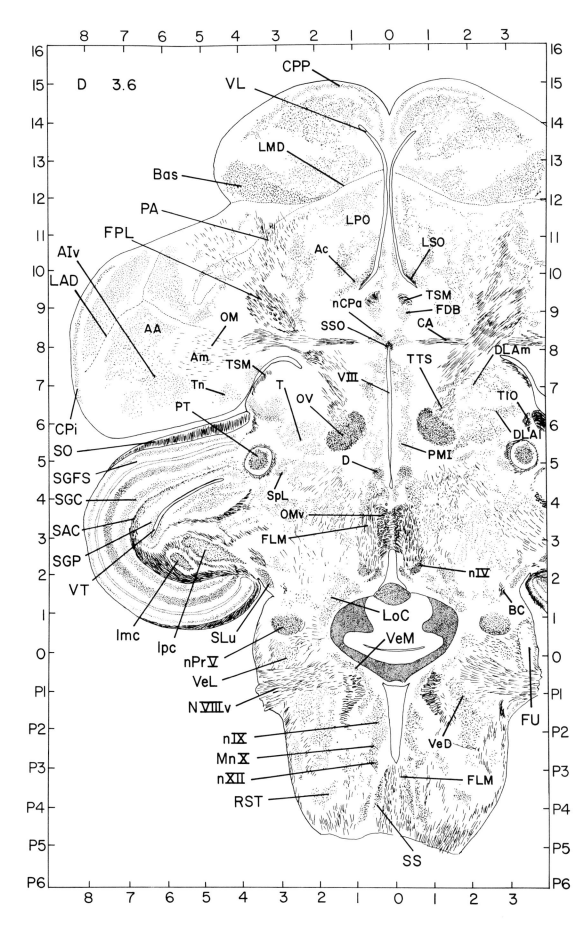

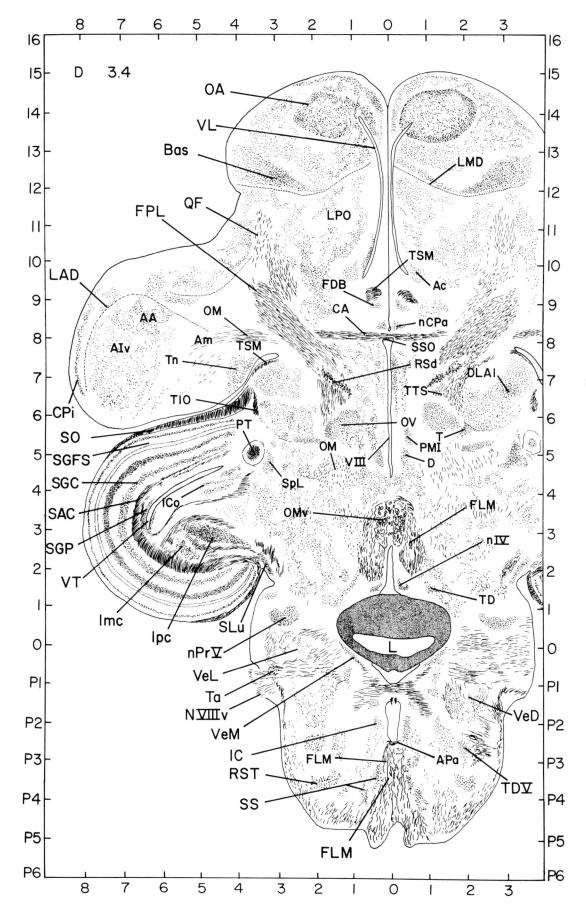

D   3.4

Ac   Nucleus accumbens
AIv   Archistriatum intermedium, pars ventralis
        (Zeier and Karten)
Am   Archistriatum mediale (Zeier and Karten)
Bas   Nucleus basalis
BC   Brachium conjunctivum
BO   Bulbus olfactorius
CA   Commissura anterior [rostralis] (Anterior
        commissure)
D   Nucleus of Darkschewitsch
DLAl Nucleus dorsolateralis anterior [rostralis]
        thalami, pars lateralis
DLAmc Nucleus dorsolateralis anterior [rostralis]
        thalami, pars magnocellularis
FDB   Fasciculus diagonalis Brocae
FLM   Fasciculus longitudinalis medialis
FPL   Fasciculus prosencephali lateralis (Lateral
        forebrain bundle)
IC   Nucleus intercalatus
ICo   Nucleus intercollicularis
Imc   Nucleus isthmi, pars magnocellularis
Ipc   Nucleus isthmi, pars parvocellularis
LAD   Lamina archistriatalis dorsalis
LMD   Lamina medullaris dorsalis
LPO   Lobus parolfactorius
Mn V Nucleus motorius nervi trigemini
nCPa Nucleus commissurae pallii (Bed nucleus
        pallial commissure)
nPr V Nucleus sensorius principalis nervi trigemini
N VIII v Nervus octavus, pars vestibularis
OA   Nucleus olfactorius anterior [rostralis]
OI   Nucleus olivaris inferior
OM   Tractus occipitomesencephalicus
OMv   Nucleus nervi oculomotorii, pars ventralis
PMI   Nucleus paramedianus internus thalami
PT   Nucleus pretectalis
PVN   Nucleus paraventricularis magnocellularis
        (Paraventricular nucleus)
QF   Tractus quintofrontalis
ROT   Nucleus rotundus
RSd   Nucleus reticularis superior, pars dorsalis
RST   Nucleus reticularis subtrigeminalis
Ru   Nucleus ruber
SAC   Stratum album centrale
SCbd Tractus spinocerebellaris dorsalis
SCv   Nucleus subceruleus ventralis
SGC   Stratum griseum centrale
SGFS Stratum griseum et fibrosum superficiale
SGP   Stratum griseum periventriculare
SLu   Nucleus semilunaris
SO   Stratum opticum
SpL   Nucleus spiriformis lateralis
SRt   Nucleus subrotundus
SS   Nucleus supraspinalis
SSO   Organum subseptale (Subseptal, subfornical
        or interventricular organ)
T   Nucleus triangularis
TD   Nucleus tegmenti dorsalis
TIO   Tractus isthmo-opticus
Tn   Nucleus taeniae
TOV   Tractus nuclei ovoidalis
TPc   Nucleus tegmenti pedunculo-pontinus, pars
        compacta (Substantia nigra)
TSM   Tractus septomesencephalicus
TTS   Tractus thalamostriaticus
TV   Nucleus tegmenti ventralis
VeD   Nucleus vestibularis descendens
VeL   Nucleus vestibularis lateralis
VeM   Nucleus vestibularis medialis
VL   Ventriculus lateralis
VT   Ventriculus tecti mesencephali
V III Ventriculus tertius (Third ventricle)

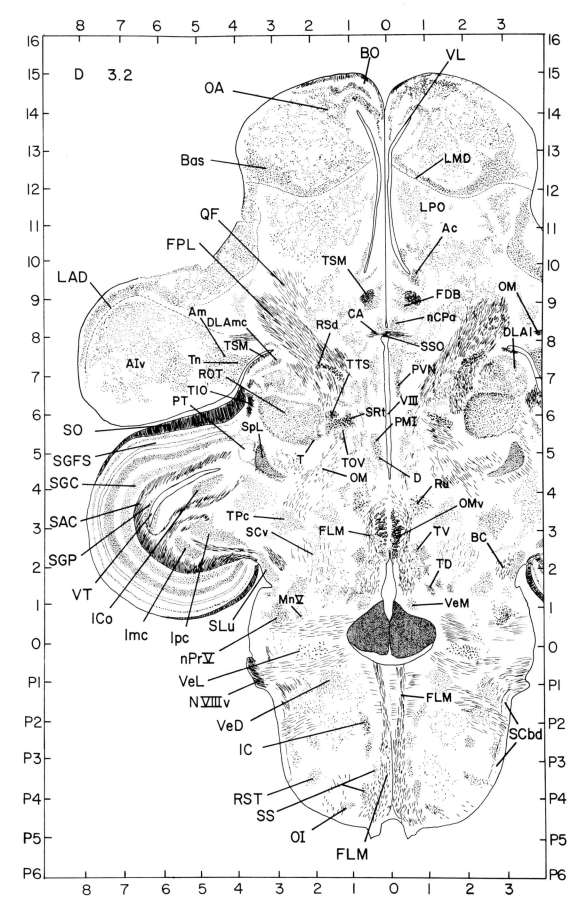

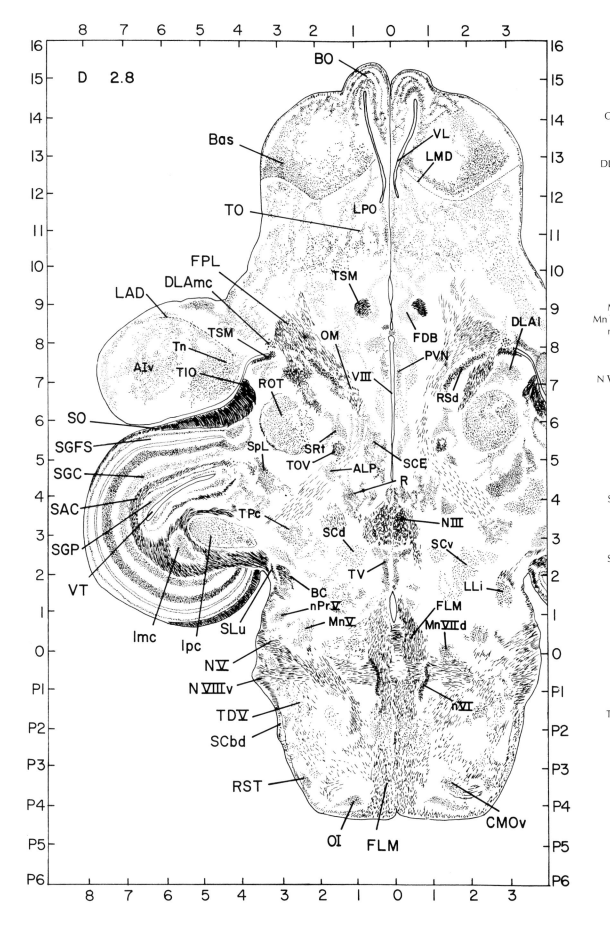

D 2.8

ALP	Nucleus ansae lenticularis posterior [caudalis]
AIv	Archistriatum intermedium, pars ventralis (Zeier and Karten)
Bas	Nucleus basalis
BC	Brachium conjunctivum
BO	Bulbus olfactorius
CMOv	Nucleus centralis medullae oblongatae, pars ventralis
DLAl	Nucleus dorsolateralis anterior [rostralis] thalami, pars lateralis
DLAmc	Nucleus dorsolateralis anterior [rostralis] thalami, pars magnocellularis
FDB	Fasciculus diagonalis Brocae
FLM	Fasciculus longitudinalis medialis
FPL	Fasciculus prosencephali lateralis (Lateral forebrain bundle)
Imc	Nucleus isthmi, pars magnocellularis
Ipc	Nucleus isthmi, pars parvocellularis
LAD	Lamina archistriatalis dorsalis
LLi	Nucleus lemnisci lateralis, pars intermedia
LMD	Lamina medullaris dorsalis
LPO	Lobus parolfactorius
Mn V	Nucleus motorius nervi trigemini
Mn VII d	Nucleus motorius nervi facialis, pars dorsalis
nPr V	Nucleus sensorius principalis nervi trigemini
n VI	Nucleus nervi abducentis
N III	Nervus oculomotorius
N V	Nervus trigeminus
N VIII v	Nervus octavus, pars vestibularis
OI	Nucleus olivaris inferior
OM	Tractus occipitomesencephalicus
PVN	Nucleus paraventricularis magnocellularis (Paraventricular nucleus)
R	Nucleus raphes
ROT	Nucleus rotundus
RSd	Nucleus reticularis superior, pars dorsalis
RST	Nucleus reticularis subtrigeminalis
SAC	Stratum album centrale
SCbd	Tractus spinocerebellaris dorsalis
SCd	Nucleus subceruleus dorsalis
SCE	Stratum cellulare externum
SCv	Nucleus subceruleus ventralis
SGC	Stratum griseum centrale
SGFS	Stratum griseum et fibrosum superficiale
SGP	Stratum griseum periventriculare
SLu	Nucleus semilunaris
SO	Stratum opticum
SpL	Nucleus spiriformis lateralis
SRt	Nucleus subrotundus
TIO	Tractus isthmo-opticus
Tn	Nucleus taeniae
TO	Tuberculum olfactorium
TOV	Tractus nuclei ovoidalis
TPc	Nucleus tegmenti pedunculo-pontinus, pars compacta (Substantia nigra)
TSM	Tractus septomesencephalicus
TD V	Nucleus et tractus descendens nervi trigemini
TV	Nucleus tegmenti ventralis
VL	Ventriculus lateralis
VT	Ventriculus tecti mesencephali
V III	Ventriculus tertius (Third ventricle)

AL Ansa lenticularis
ALP Nucleus ansae lenticularis posterior [caudalis]
AIv Archistriatum intermedium, pars ventralis (Zeier and Karten)
Bas Nucleus basalis
BO Bulbus olfactorius
FDB Fasciculus diagonalis Brocae
FLM Fasciculus longitudinalis medialis
FPL Fasciculus prosencephali lateralis (Lateral forebrain bundle)
Imc Nucleus isthmi, pars magnocellularis
Ipc Nucleus isthmi, pars parvocellularis
LA Nucleus lateralis anterior [rostralis] thalami
LC Nucleus linearis caudalis
LLi Nucleus lemnisci lateralis, pars intermedia
LLv Nucleus lemnisci lateralis, pars ventralis
LMD Lamina medullaris dorsalis
LMmc Nucleus lentiformis mesencephali, pars magnocellularis
LMpc Nucleus lentiformis mesencephali, pars parvocellularis
Mn V Nucleus motorius nervi trigemini
Mn VII d Nucleus motorius nervi facialis, pars dorsalis
MPv Nucleus mesencephalicus profundus, pars ventralis (Jungherr)
N III Nervus oculomotorius
N V Nervus trigeminus
n VI Nucleus nervi abducentis
OI Nucleus olivaris inferior
OS Nucleus olivaris superior
POM Nucleus preopticus medialis
PPC Nucleus principalis precommissuralis
PVN Nucleus paraventricularis magnocellularis (Paraventricular nucleus)
R Nucleus raphes
Rgc Nucleus reticularis gigantocellularis
RPgc Nucleus reticularis pontis caudalis, pars gigantocellularis
ROT Nucleus rotundus
RPO Nucleus reticularis pontis oralis
RSd Nucleus reticularis superior, pars dorsalis
Ru Nucleus ruber
SAC Stratum album centrale
SCbd Tractus spinocerebellaris dorsalis
SCE Stratum cellulare externum
SGC Stratum griseum centrale
SGFS Stratum griseum et fibrosum superficiale
SGP Stratum griseum periventriculare
SO Stratum opticum
SOv Nucleus supraopticus, pars ventralis
SpL Nucleus spiriformis lateralis
RST Nucleus reticularis subtrigeminalis
TO Tuberculum olfactorium
TOV Tractus nuclei ovoidalis
TPc Nucleus tegmenti pedunculo-pontinus, pars compacta (Substantia nigra)
TSM Tractus septomesencephalicus
TD V Nucleus et tractus descendens nervi trigemini
TV Nucleus tegmenti ventralis
VO Ventriculus olfactorius
VT Ventriculus tecti mesencephali
V III Ventriculus tertius (Third ventricle)

D 2.4

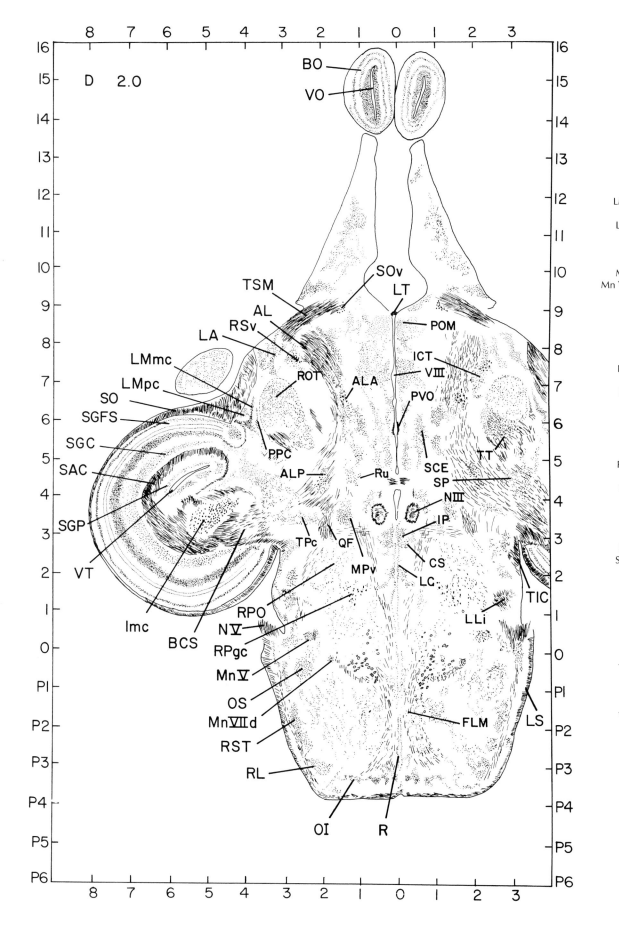

AL   Ansa lenticularis
ALA   Nucleus ansae lenticularis anterior [rostralis]
ALP   Nucleus ansae lenticularis posterior [caudalis]
BCS   Brachium colliculi superioris
BO   Bulbus olfactorius
CS   Nucleus centralis superior
FLM   Fasciculus longitudinalis medialis
ICT   Nucleus intercalatus thalami
Imc   Nucleus isthmi, pars magnocellularis
IP   Nucleus interpeduncularis
LA   Nucleus lateralis anterior [rostralis] thalami
LC   Nucleus linearis caudalis
LLi   Nucleus lemnisci lateralis, pars intermedia
LMmc   Nucleus lentiformis mesencephali, pars magnocellularis
LMpc   Nucleus lentiformis mesencephali, pars parvocellularis
LS   Lemniscus spinalis
LT   Lamina terminalis
Mn V   Nucleus motorius nervi trigemini
Mn VII d   Nucleus motorius nervi facialis, pars dorsalis
MPv   Nucleus mesencephalicus profundus, pars ventralis (Jungherr)
N III   Nervus oculomotorius
N V   Nervus trigeminus
OI   Nucleus olivaris inferior
OS   Nucleus olivaris superior
POM   Nucleus preopticus medialis
PPC   Nucleus principalis precommissuralis
PVO   Organum paraventriculare (Paraventricular organ)
QF   Tractus quintofrontalis
R   Nucleus raphes
RL   Nucleus reticularis lateralis
ROT   Nucleus rotundus
RPgc   Nucleus reticularis pontis caudalis, pars gigantocellularis
RPO   Nucleus reticularis pontis oralis
RSv   Nucleus reticularis superior, pars ventralis
Ru   Nucleus ruber
SAC   Stratum album centrale
SCE   Stratum cellulare externum
SGC   Stratum griseum centrale
SGFS   Stratum griseum et fibrosum superficiale
SGP   Stratum griseum periventriculare
SO   Stratum opticum
SOv   Nucleus supraopticus, pars ventralis
SP   Nucleus subpretectalis
RST   Nucleus reticularis subtrigeminalis
TIC   Tractus isthmocerebellaris
TPc   Nucleus tegmenti pedunculo-pontinus, pars compacta (Substantia nigra)
TSM   Tractus septomesencephalicus
TT   Tractus tectothalamicus
VO   Ventriculus olfactorius
VT   Ventriculus tecti mesencephali
V III   Ventriculus tertius (Third ventricle)

AL   Ansa lenticularis
AM   Nucleus anterior [rostralis] medialis
       hypothalami
AVT Area ventralis (Tsai)
BCD Brachium conjunctivum descendens
BCS Brachium colliculi superioris
BO   Bulbus olfactorius
CS   Nucleus centrális superior
DSM Decussatio supramamillaris
GLv Nucleus geniculatus lateralis, pars ventralis
Imc Nucleus isthmi, pars magnocellularis
ICT Nucleus intercalatus thalami
IP   Nucleus interpeduncularis
LC   Nucleus linearis caudalis
LMmc Nucleus lentiformis mesencephali, pars
       magnocellularis
LMpc Nucleus lentiformis mesencephali, pars
       parvocellularis
LS   Lemniscus spinalis
LT   Lamina terminalis
MPv Nucleus mesencephalicus profundus, pars
       ventralis
Mn VII v Nucleus motorius nervi facialis, pars
       ventralis
N III Nervus oculomotorius
nI   Nucleus intramedialis
OI   Nucleus olivaris inferior
OS   Nucleus olivaris superior
Pap Nucleus papillioformis
Rpgl Nucleus reticularis paragigantocellularis
       lateralis
PHN Nucleus periventricularis hypothalami
PL   Nucleus pontis lateralis
POM Nucleus preopticus medialis
PPC Nucleus principalis precommissuralis
PVO Organum paraventriculare (Paraventricular
       organ)
R   Nucleus raphes
ROT Nucleus rotundus
Rpc Nucleus reticularis parvocellularis
RPgc Nucleus reticularis pontis caudalis, pars
       gigantocellularis
RPO Nucleus reticularis pontis oralis
SAC Stratum album centrale
SCv Nucleus subceruleus ventralis
SGC Stratum griseum centrale
SGFS Stratum griseum et fibrosum superficiale
SO   Stratum opticum
SOv Nucleus supraopticus, pars ventralis
SP   Nucleus subpretectalis
TT   Tractus tectothalamicus
VLT Nucleus ventrolateralis thalami
VO   Ventriculus olfactorius
V III Ventriculus tertius (Third ventricle)

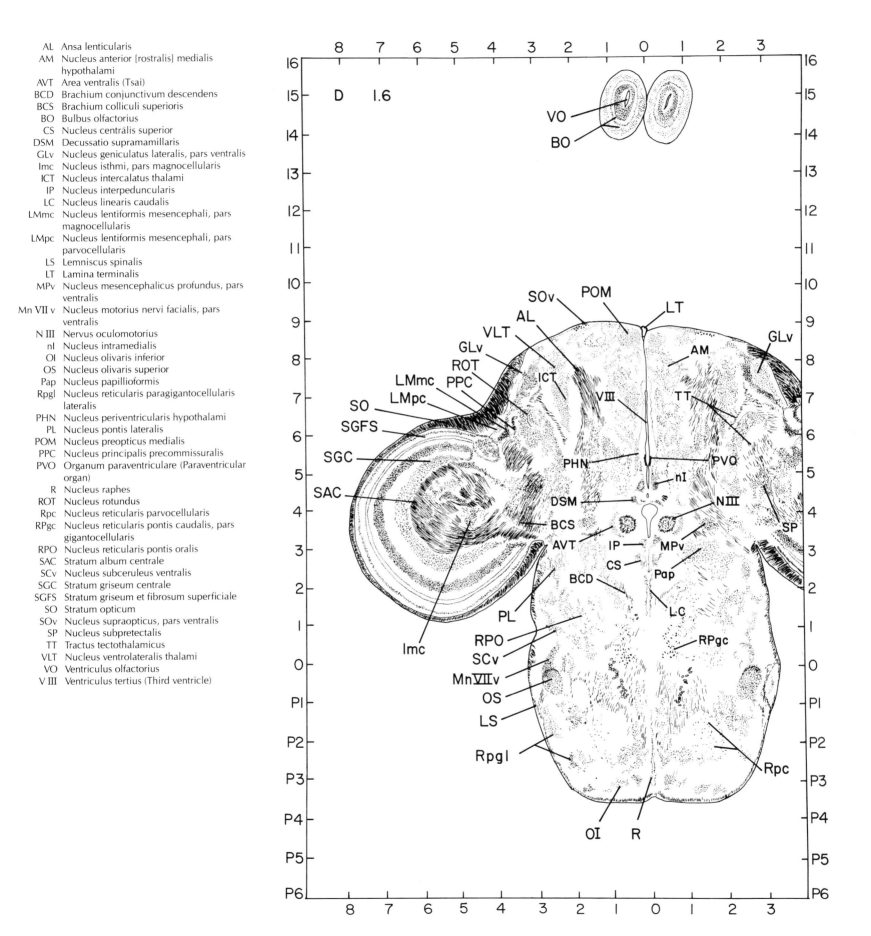

D   1.6

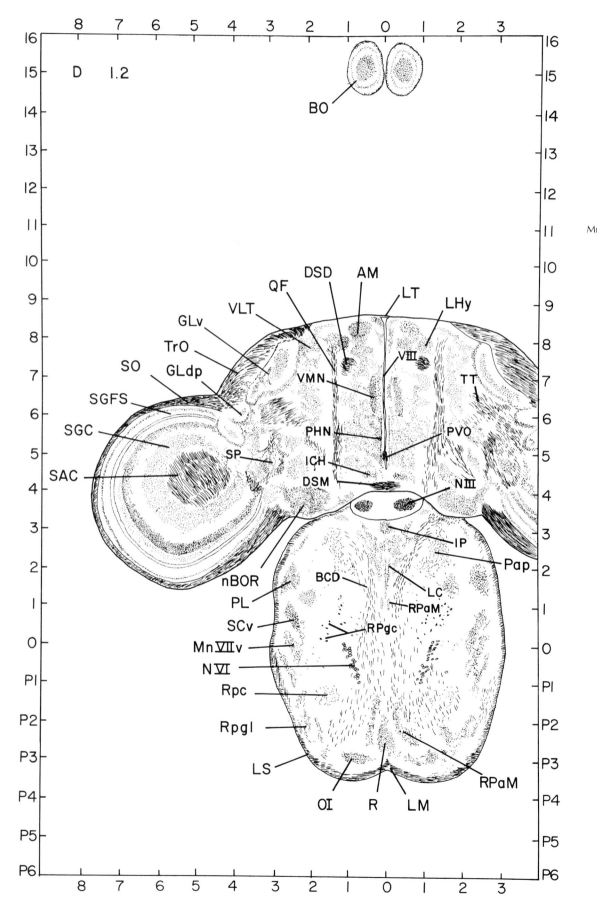

AM   Nucleus anterior [rostralis] medialis hypothalami
DMN   Nucleus dorsomedialis hypothalami
DSD   Decussatio supraoptica dorsalis
DSV   Decussatio supraoptica ventralis
GLdp   Nucleus geniculatus lateralis, pars dorsalis principalis
GLv   Nucleus geniculatus lateralis, pars ventralis
IP   Nucleus interpeduncularis
LHy   Regio lateralis hypothalami (Lateral hypothalamic area)
LM   Lemniscus medialis
LS   Lemniscus spinalis
LT   Lamina terminalis
nBOR   Nucleus opticus basalis; nucleus ectomamillaris (Nucleus of the basal optic root)
N VI   Nervus abducens
Pap   Nucleus papillioformis
PHN   Nucleus periventricularis hypothalami
PL   Nucleus pontis lateralis
PVO   Organum paraventriculare (Paraventricular organ)
R   Nucleus raphes
RPaM   Nucleus reticularis paramedianus
RPR   Recessus preopticus
SGC   Stratum griseum centrale
SGFS   Stratum griseum et fibrosum superficiale
SO   Stratum opticum
TrO   Tractus opticus
VMN   Nucleus ventromedialis hypothalami
V III   Ventriculus tertius (Third ventricle)

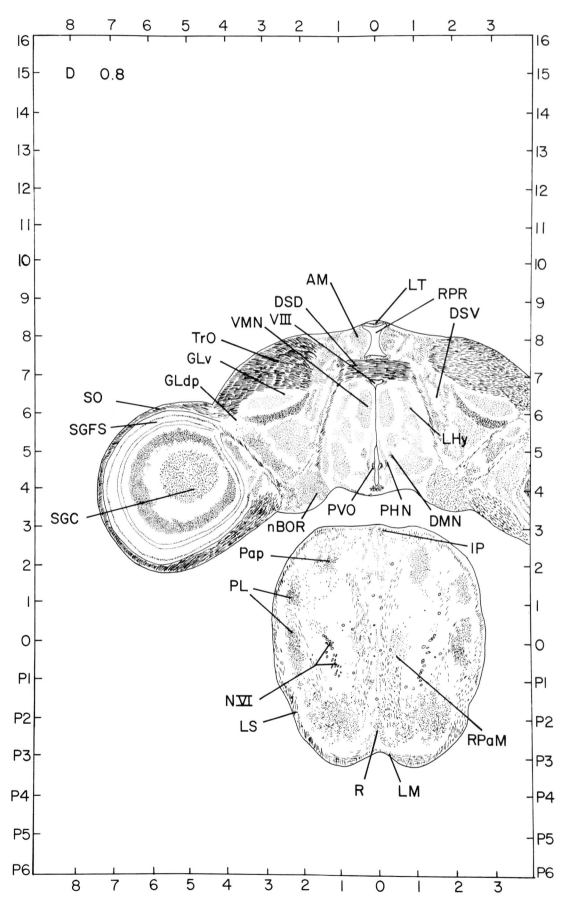

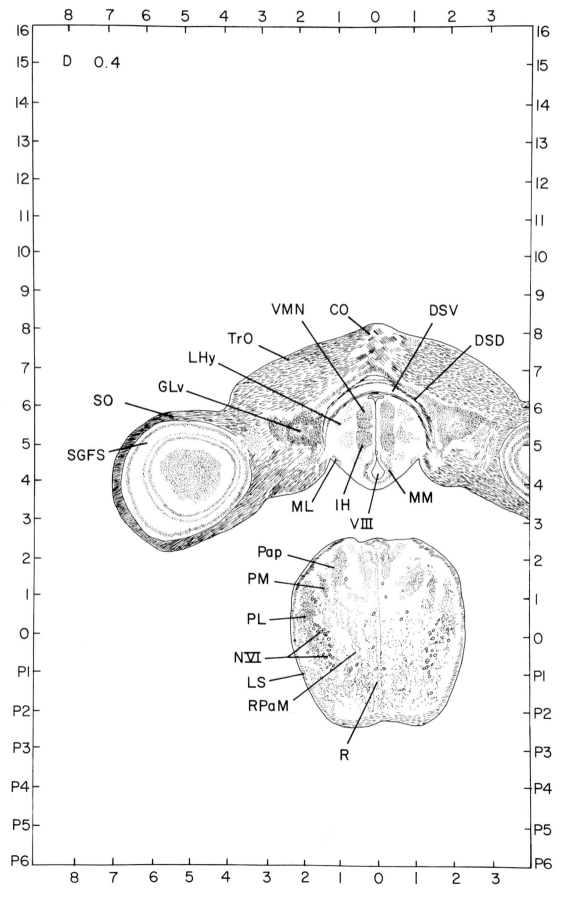

D 0.4

CO   Chiasma opticum
DSD   Decussatio supraoptica dorsalis
DSV   Decussatio supraoptica ventralis
GLv   Nucleus geniculatus lateralis, pars ventralis
IH   Nucleus inferioris hypothalami
LHy   Regio lateralis hypothalami (Lateral
      hypothalamic area)
LS   Lemniscus spinalis
ML   Nucleus mamillaris lateralis
MM   Nucleus mamillaris medialis
N VI Nervus abducens
Pap   Nucleus papillioformis
PL   Nucleus pontis lateralis
PM   Nucleus pontis medialis
R   Nucleus raphes
RPaM Nucleus reticularis paramedianus
SGFS Stratum griseum et fibrosum superficiale
SO   Stratum opticum
TrO   Tractus opticus
VMN   Nucleus ventromedialis hypothalami
V III Ventriculus tertius (Third ventricle)

CO   Chiasma opticum
CTz   Corpus trapezoideum
IH   Nucleus inferioris hypothalami
MM   Nucleus mamillaris medialis
N VI Nervus abducens
PM   Nucleus pontis medialis
R   Nucleus raphes
SGFS Stratum griseum et fibrosum superficiale
SO   Stratum opticum
TrO   Tractus opticus
VMN   Nucleus ventromedialis hypothalami
V III Ventriculus tertius (Third ventricle)

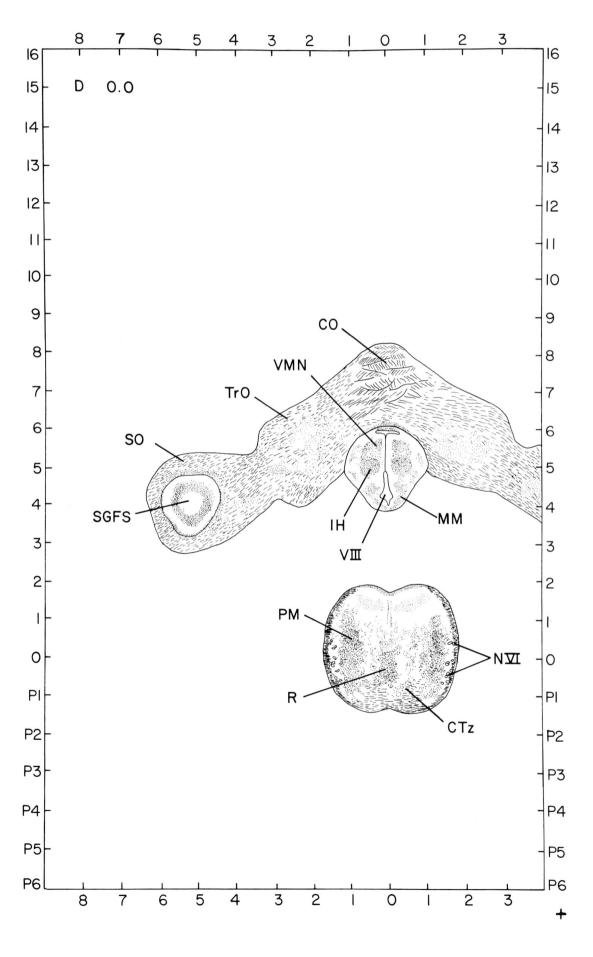

D 0.0

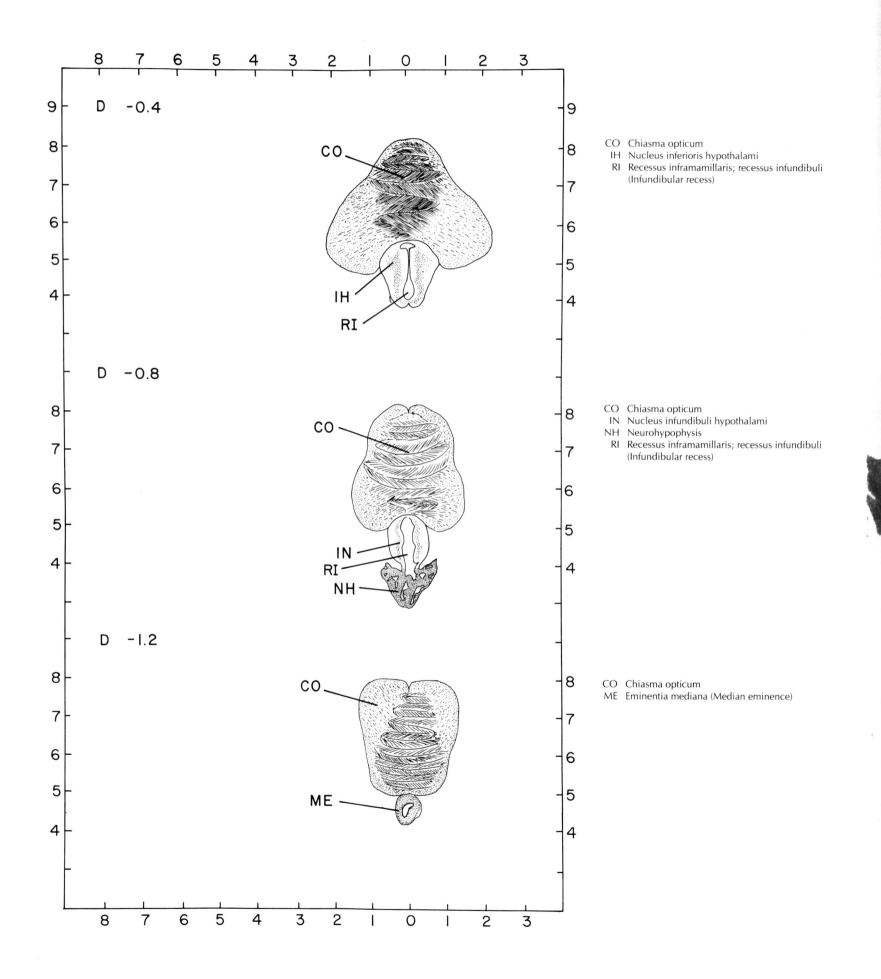

D -0.4

CO   Chiasma opticum
 IH   Nucleus inferioris hypothalami
 RI   Recessus inframamillaris; recessus infundibuli
        (Infundibular recess)

D -0.8

CO   Chiasma opticum
 IN   Nucleus infundibuli hypothalami
NH   Neurohypophysis
 RI   Recessus inframamillaris; recessus infundibuli
        (Infundibular recess)

D -1.2

CO   Chiasma opticum
ME   Eminentia mediana (Median eminence)